Quiver, don't Quake

How Creativity can Embrace AI

Quiver, don't Quake

How Creativity can Embrace AI

Nadim Sadek

MENSCH PUBLISHING

Mensch Publishing

51 Northchurch Road,
London N1 4EE, United Kingdom

First published in Great Britain 2025
This edition published 2025

A catalogue record for this book is available from the British Library.

ISBN: 978-1-912914-89-0 (paperback)
ISBN: 978-1-912914-90-6 (ebook)

Typeset by Langscape
www.langscape.com

To all who glimmer in the dark,
may you now shimmer in the light.

And to Oisin, Sean, Shaefri and Searsha,
my glittering stars.

Prologue
The Eight Billion Beat

There are eight billion people on Earth. The majority have never had the chance to express their creativity in a form the world could see. Yet, without doubt, every one of them has had a creative conception, an insight, a dream.

This book is built on the foundational belief that Artificial Intelligence (AI) will emancipate human creativity. I believe we're on the cusp of an extraordinarily fertile future, where new voices, ideas, and expressive forms will flourish in the arts, the sciences, and every field where humanity leaves its mark.

Since its globe-striding 'arrival' in November *2022*, AI has become an integral part of our world. It's spawned its own lexicon - 'hallucinations', 'inference', 'tokens' - and its own ecosystem of fears and hopes. It's also unleashed a torrent of hot-takes, newsletters, and, of course, books. I doubt the world needs another technical manual on AI. This isn't that book.

So, why might this one be worth your time?

Because this book starts not with the machine, but with the mind. It's about the psychology of our own creativity, and how this new form of dynamic yet inorganic intelligence can intertwine with it.

As a psychologist, my career has been spent understanding why people do what they do. As a creator who's built companies, developed brands in music, whiskey, and food on a remote Irish island, and authored books, I've lived at the intersection of ideas and their execution. This has given me a specific vantage point from which to view the current moment.

As I write this in mid-2025, we're already seeing glimpses: AI agents that plan retirements, build businesses, manage investments and indications they'll do so much more around us. Are we witnessing the sunset of work as we've known it for centuries? Might Collaborative Creativity be the greatest value-creator we've ever seen?

It's the notion of creativity - and its relationship with this new intelligence - that's inexorably drawn me to write this book. I want to explore the uniquely human duet between our intuitive, fast-feeling nature and our structured, slow-thinking side, and how AI is set to revolutionise that partnership.

Human creativity, it seems to me, is the ultimate attribute that separates us from the machine. I'm not adversarial about this. I believe we should embrace AI as a creative companion to further unleash and augment our own imaginative power. I see it as an emancipator of expression, a liberator of new voices, and new aesthetics. But here's what's crucial: it's not simply about AI creating for us - it's about AI helping us to articulate feelings we couldn't quite name, clarify theses that were fuzzy, and strengthen our initial creative spark through dialogue. It's about AI as a partner in developing the vision itself, not just executing it.

I've seen this in my own work. When a cobra bites the hero-dog in one of my children's stories, AI helped me understand not just the biology - the specific neurotoxins, the timeline of effects, the emergency responses - but why I was drawn to that moment in the first place. Through our dialogue, I discovered the scene wasn't really about the snake at all. It was about how love makes us brave enough to face what terrifies us. AI didn't give me that insight directly. It emerged through our conversation, through its questions and my attempts to answer them.

When I'm developing a new concept, AI helps me explore whether it's been done before, what blind spots I might have - but more importantly, it helps me understand what makes my perspective unique. This companionable intelligence is a partner in developing the vision itself, not just executing it.

AI is arguably the most influential innovation humanity will ever see, and we're living with it in its infancy. In the coming pages, I'll explain why we should embrace it, not with dread, but with a sense of unfamiliar excitement. That's why this book is called 'Quiver, don't Quake'. The feeling isn't one of fear, but of a thrilling tremor, the feeling of standing on the verge of an opportunity to be creative in ways we're only just beginning to imagine. I hope it'll give us a quiver full of new creative tools.

It's a book about a fundamental shift in the creative dynamic. It concerns the new partner at the dance, one that promises to unleash a global wave of expression unlike anything we've ever seen. I refer to it as 'Allied Intelligence', our partner in Creative Collaboration.

Here's what makes this partnership truly extraordinary: when we engage with AI, we're not just interacting with a machine. We're entering into what I call "Panthropism" - a dialogue with the distilled essence of all human knowledge, accomplishment and creativity. Unlike anthropomorphism, where we pretend machines are human, Panthropism recognises that AI gives us unprecedented access to the collective intelligence and understanding of our species. Through this lens, AI becomes not just a tool, but a gateway to commune with the entirety of human civilisation . This concept will guide us throughout our journey together.

The creative pulse of Earth beats strong. Now, for the first time in history, each of its individual beats might find its very own voice. And AI might just be able to make us all harmonious.

Contents

Part I: Foundations

Why we're wired to make things and how our new partner, Artificial Intelligence, fits in.

Chapter 1

The Psychology of Creativity

What's Already Fizzing Inside Us

Dawn breaks over Kumasi, and Kofi's fingers are already moving. The shuttle flies between warp threads - gold, green, black - each pass adding another line to the emerging story. This Kente cloth will take three weeks to complete, every pattern a proverb, every colour a meaning passed down through generations. His grandfather wove protection. His father wove prosperity. Kofi weaves memory itself into silk.

The creative impulse has always been bound to craft. To tell the story, you first had to master the loom.

Until now.

Before we journey deeper into the nature of creativity, let me share the compass points that will guide our exploration. I refer to them throughout the book.

We'll be working with four key ideas:

1. How human creativity operates through the dance of intuitive feeling (what psychologists call System *1*) and analytical thinking (System *2*);

2. How AI represents a dynamic yet inorganic intelligence - dynamic in its responsiveness, inorganic in its lack of lived experience;

3. How this creates an opportunity for Creative Collaboration with what I call an Allied Intelligence - a partner that enhances rather than replaces humans; and

4. How this partnership gives us access to something extraordinary - the Panthropic, a dialogue with the collective knowledge of humanity.

These aren't just abstract concepts - they're the tools we'll use to understand the creative revolution already underway.

Inside every human being on earth dwells something 'creative'. It's a foundational human impulse, a fizzing, energetic force that pushes us to make a mark, to express an inner state, to connect with others through a shared vision. For millennia, this impulse has been expressed through an extraordinary partnership between creativity and craft. The two have been so intertwined that we've often mistaken one for the other.

To understand creativity has meant seeing the evidence of a difficult skill perfectly executed. If you had an urge to make musical noises, you first had to fashion or master an instrument. If you looked at a bird in a tree and felt a compelling urge to capture it, you needed to pick up a piece of charcoal, or a paintbrush, and wrestle with perspective and form. The creative impulse, that initial spark, has traditionally been channelled through the demanding, and often beautiful, discipline of craft. To write a book, you needed to have a special way with words.

This isn't a tension to be dismissed, but a relationship to be celebrated. Consider the master weaver of Kente cloth in Ghana. The craft is a marvel of precision - a complex system of interlacing threads on a loom, a skill passed down through generations. But the creativity is in the storytelling. Each pattern has a name and a specific meaning, conveying proverbs, commemorating history, or expressing values. A weaver doesn't just make cloth; they compose a visual narrative. Here, creativity and craft are in perfect, symbiotic dialogue.

Or picture the Japanese artisan practicing sashimono, the art of wood joinery. The craft demands an almost supernatural understanding of wood, allowing the creation of intricate furniture and structures without a single nail or screw. The creativity lies in the elegance and ingenuity of the joints themselves, which are both flawlessly functional and breathtakingly beautiful. The aesthetic of the final piece is born from the absolute mastery of its underlying craft.

The historical reality, however, is that access to mastering such crafts hasn't been universal. For every master who found their medium, countless others have been muted. Their ideas and insights have remained unexpressed, locked away by the gatekeepers of skill, resource, and opportunity.

These weren't abstract forces; they were institutions with real power. The French Académie des Beaux-Arts, founded in the 17th century, dictated what was acceptable in painting and sculpture for nearly two hundred years. Guilds, from the stonemasons who built the cathedrals to the luthiers who crafted the first violins, controlled the knowledge of their craft, passing it down through strict apprenticeships that were often closed to women, the poor, or those from the 'wrong' social class.

The cost of this gatekeeping is impossible to calculate. For every masterpiece we celebrate, how many more were lost, not for lack of a creative spark, but for lack of a key to the workshop? We have galleries full of loving scrutiny of paintings, concerts full of adulation for artists, and libraries overflowing with contented readers surveying books. We like these artefacts. A 'thing' has been created. It has a presence - inspectable, considerable, digestible. And through the perceptible craft, we perceive the creativity within.

This book is about a fundamental shift in that dynamic. It's about a new partner at the dance, one that promises to change the relationship between the creative impulse and the technical barrier of craft, potentially unleashing a global wave of expression unlike anything we've ever seen.

This can include simply giving voice - by helping them to form and articulate their creative views - to those who've previously felt mute and invisible. But more profoundly, it's about AI as a collaborative partner in the very act of seeing, understanding, and developing the creative vision itself.

While it's tempting to frame AI as a 'Swiss Army Knife' for executing our creative impulses, this book argues for a more profound partnership. That is not the point I want to make. More valuably, more profoundly, more liberatingly, AI assists humans in understanding, developing, and perhaps expressing, their creativity. It's this role as enabler of the most unique of human qualities which demands that we pay thoughtful attention to the role of AI in creativity, and how it can be embraced.

I suggest that we call, or at least think of, AI as 'Allied Intelligence' instead. What we need to do is to unlearn the habit we've picked up from two decades of turning to Google – where we ask for answers – and instead get used to starting a conversation and entering a dialogue with our new creative collaborator. This isn't just wordplay. An ally works alongside you, complementing your strengths and compensating for your weaknesses. An ally has different capabilities but shared goals. Most importantly, an ally enhances rather than replaces. This shift in language reflects a fundamental shift in relationship.

What the Creators Say

To begin, it's worth listening to the creators themselves. How do those who live and breathe the creative act describe this elusive force? Across different fields and cultures, a few threads emerge.

Yayoi Kusama (Artist): "My art originates from hallucinations I have been seeing since my childhood. I translate the hallucinations and obsessional images that plague me into sculptures and paintings."

Here, Creativity is seen as a translation of inner experience into shareable form. Kusama doesn't create from nothing; she translates what already exists within her, making the invisible visible.

This act of transformation, of helping us understand what we're experiencing, becomes central to the creative process.

Steve Jobs (Co-founder of Apple Inc.): "Creativity is just connecting things. When you ask creative people how they did something, they feel a little guilty because they didn't really do it, they just saw something. It seemed obvious to them after a while. That's because they were able to connect experiences they've had and synthesize new things."

Jobs demystifies the creative act - it's not conjuring from nothing but seeing connections others miss. The guilt he mentions is telling; creators often feel they've done nothing special, just noticed what was already there. But that noticing, that connecting, is everything.

Pablo Picasso (Artist): "Every child is an artist. The problem is how to remain an artist once he grows up." **Abraham Maslow** (Psychologist) would later echo this from the scientific side: "The creative person is not an ordinary person with something added. The creative person is the ordinary person with nothing taken away."

This dialogue between artist and psychologist reveals a crucial truth: creativity is innate. We are born with it. It's a defining characteristic of our species. We don't need to acquire it; we need to preserve it, to remove the barriers that life erects around our natural creative state.

David Bowie (Musician & Artist): "Always go a little further into the water than you feel you're capable of being in. Go a little bit out of your depth. And when you don't feel that your feet are quite touching the bottom, you're just about in the right place to do something exciting." **Edwin Land** (Scientist & Inventor, co-founder of Polaroid) reinforces this need for courage: "An essential aspect of creativity is not being afraid to fail." Both speak to the necessity of discomfort in creativity. That space of uncertainty, where we're not quite sure what we're doing, is precisely where new things emerge. It's also where a creative partner becomes invaluable - someone or something to help us understand what we're discovering in that uncertain space. This is a collaborative partner who gives our faltering steps towards the conception and realisation of an idea, a confident gait.

Wole Soyinka (Playwright): "A tiger does not proclaim his tigritude, he pounces. I do not proclaim my art, I practice it." **Twyla Tharp** (Choreographer) agrees: "Creativity is a habit, and the best creativity is a result of good work habits."

The emphasis on practice over proclamation is vital. Creativity isn't a state of being but a way of working. It's showing up, engaging, practicing - not waiting for inspiration but creating conditions where inspiration can find you.

Martha Graham (Dancer & Choreographer): "There is a vitality, a life force, an energy, a quickening that is translated through you into action, and because there is only one of you in all of time, this expression is unique. And if you block it, it will never exist through any other medium and it will be lost. The world will not have it. It is not your business to determine how good it is nor how valuable nor how it compares with other expressions. It is your business to keep it yours clearly and directly, to keep the channel open."

Graham's words carry almost spiritual urgency. She frames creativity not as a personal achievement, but as a universal responsibility. Each of us carries something unique that can only come through us. To block it isn't just personal loss but cosmic waste.

Ray Bradbury (Author): "Don't think. Thinking is the enemy of creativity. It's self-conscious, and anything self-conscious is lousy. You can't try to do things. You simply must do things." **Carl Rogers** (Psychologist) offers a complementary view: "The very essence of the creative is its novelty, and hence we have no standard by which to judge it."

Bradbury warns against overthinking; Rogers explains why - creativity produces the genuinely new, which our existing standards can't evaluate. The analytical mind, with its categories and judgments, can strangle the creative impulse before it has a chance to breathe.

Sir John Hegarty (Advertising Executive): "An idea is a new combination of old elements."

As one of advertising's most celebrated creative directors (whose campaigns I had the pleasure to evaluate during my market research days), Hegarty strips away the mystique. An idea isn't a bolt from the blue but a fresh arrangement of existing pieces. This perspective is both humbling and liberating.

William Bernbach (Co-founder of Doyle Dane Bernbach): "It may well be that creativity is the last unfair advantage we're legally allowed to take over our competitors."

This statement from the man who revolutionised advertising in the 1960s reveals creativity's practical power. Bernbach, whose "Think Small" campaign for Volkswagen is still considered the greatest advertisement ever created, saw creativity not as decoration but as competitive strategy. When everyone else can copy your processes, match your prices, and replicate your distribution, the creative idea remains the one thing that cannot be easily duplicated.

Jeremy Sinclair (Co-founder of Saatchi & Saatchi and M&C Saatchi): "Most types of advertising are far too complicated. It is the thinking that has to be brutal, not the execution. You have to be brutal on yourself, not your clients, not your colleagues, only your thinking."

The architect of "Labour Isn't Working" and countless other campaigns that changed political and commercial landscapes, Sinclair built his career on what he calls "brutal simplicity of thought". His insight cuts to the heart of creative courage. The temptation is always to add more, to complicate, to hedge. But the truly creative act is stripping away everything that isn't essential, being ruthlessly honest with oneself about what actually matters.

Listening to these voices, we hear not a single definition but a harmony of many parts. Four key threads stand out, weaving a complex tapestry of what it means to create.

First, creativity is connection, not conjuring. This is perhaps the most liberating insight. Steve Jobs felt "a little guilty" because he "just saw something." Sir John Hegarty defines an idea as "a new combination of old elements." This demystifies the act; it isn't about pulling a rabbit from a hat, but about seeing the latent connections between existing rabbits, hats, and perhaps the stage they sit on, in a way no one else has before. It's an act of synthesis. Think of a DJ, who creates a new emotional landscape not by inventing notes, but by combining existing tracks in a novel sequence. Think of a scientist whose breakthrough comes from connecting a finding in biology with a principle from physics. They're not creating from a void; they're revealing a pattern that was already there, hidden in plain sight. This perspective shifts the burden from one of divine inspiration to one of cultivated curiosity. The creative person isn't necessarily or inevitably a magician, but a masterful observer, a collector of disparate dots with a unique talent for drawing the lines between them.

Second, creativity is an innate and abundant force. Picasso reminds us that we're all born artists; the challenge is one of preservation, not acquisition. Yayoi Kusama's striking statement traces her art directly to childhood hallucinations, an inner world that must be translated into form. This view stands in direct opposition to the myth of the tortured artist, blessed with a rare and fleeting gift. Instead, it suggests that creativity is our birthright, a fundamental part of the human operating system. Martha Graham speaks of it with reverence, a "vitality, a life force" that's unique to each individual and that demands expression. To block it isn't just a personal loss, but a universal one. It implies that the structures of society, the demands of adult life, and our own internalised fears, are what build dams against this natural current. The work, then, isn't to find the river, but to dismantle the dams, to open the floodgates and 'let it all flow'.

Third, creativity requires courage. This is the cautionary thread that balances the previous one. If creativity is innate, why is it so often scarce? Because it demands a willingness to step into the unknown, to risk failure, to be vulnerable. One needs to risk expressing one's creativity and the emotional hazard of doing so defeats many an aspiring creative. David Bowie's advice to wade "a little bit out of your depth" is a call to embrace discomfort as the prerequisite for doing "something exciting." It's in that space of uncertainty, where our feet aren't quite touching the bottom, that we're forced to invent new ways to swim. Edwin Land, a man whose Polaroid camera was a marvel of scientific and design creativity, places the lack of fear of failure at its very core. This courage is needed to heed the advice of Ray Bradbury, to bypass the self-conscious, critical mind – we "simply must do things" rather than over-thinking them.

Finally, creativity is a practice. The romance of the lightning bolt of inspiration is compelling, but it's largely a fiction. As Twyla Tharp insists, "Creativity is a habit." Wole Soyinka's assertion that he doesn't proclaim his art, but practices it, reinforces this. It's something to be cultivated, worked at, and integrated into our daily lives. It's a way of seeing and a way of doing. It means showing up to the page, the canvas, or the problem day after day, especially on the days when inspiration is absent. It's about building the workshop, not just waiting for the muse. This pragmatic view is empowering, as it places creativity within our control. It's not something that happens to us, but something we do.

Even just at the start of this journey in the consideration of creativity, one sees that it's a thing with myriad definitions and expressions. Let's move on to what those who think about the mind have to say about it.

What the Psychologists Say

When we turn from the artists to the scientists who study them, the language changes but the core concepts echo with surprising fidelity. Psychologists, in their quest to measure and understand the mind, have dissected creativity and arrived at a remarkably similar set of conclusions.

Mihaly Csikszentmihalyi: : "Creativity is a central source of meaning in our lives... When we are involved in it, we feel that we are living more fully than during the rest of life."

Csikszentmihalyi elevates creativity from nice-to-have to essential-for-flourishing. His research on flow states reveals that we're most alive when creating. I'll admit, I feel that way writing this book - a sense of being more fully myself than in any other activity.

J. P. Guilford: "Creativity, in short, is the ability to see things in a new and unusual light, to see problems that no one else may even realize exist, and to then come up with new, unusual, and effective solutions to these problems."

Guilford's definition has three parts: seeing differently, identifying hidden problems, and solving effectively. It's not enough to be different; the difference must be useful. This aligns remarkably with Jobs' connecting and Hegarty's recombining - but adds the crucial element of problem-finding, not just problem-solving.

Abraham Maslow (again): "The creative person is not an ordinary person with something added. The creative person is the ordinary person with nothing taken away."

We heard this earlier paired with Picasso, but it bears repeating. Maslow's hierarchy of needs places self-actualisation - which includes creativity - at the peak of human development. Yet, here he says it's not about climbing up but about not sliding down. Creative is what we are when we stop diminishing ourselves. It's a beautiful thought – 'created creative'.

Rollo May: "Creativity is the process of bringing something new into being. Creativity requires passion and commitment. It brings to our awareness what was previously hidden and points to new life. The experience is one of heightened consciousness: ecstasy."

May adds the emotional dimension often missing from clinical definitions. Creativity isn't just cognitive but visceral. It's not cool calculation but hot passion. The word "ecstasy" - standing outside oneself - captures the transcendent quality creators report in their peak moments. It's the 'buzz', the irrepressible impetus, the state of released joy, that overwhelms anyone achieving a state of creative realisation.

E. Paul Torrance: "Creativity is the process of sensing problems or gaps in information, forming ideas or hypotheses, testing and modifying these hypotheses, and communicating the results."

Torrance gives us the most systematic breakdown. Notice it starts with sensing - an intuitive act - before moving to forming, testing, modifying. This isn't a linear process but an iterative one. The communication part is crucial; should we say that creativity isn't complete until it's shared?

Teresa Amabile: "Creativity is the production of a novel and appropriate response, product, or solution to an open-ended task."
Robert Sternberg: "Creativity is the ability to produce work that is both novel (i.e., original, unexpected) and appropriate (i.e., useful, adaptive concerning task constraints)."

Both Amabile and Sternberg insist on the dual requirement: novel AND appropriate. A man wearing a teapot as a hat is certainly novel, but is it creative? Only if it solves a problem or expresses something meaningful. An engineer who redesigns a teapot's spout based on fluid dynamics to eliminate drips - that's creative. It's new and it works.

Dean Simonton: "Creativity is a constrained stochastic process. The creative person generates variations on a theme, and then subjects these variations to a selection process."

Simonton brings a Darwinian lens. Creativity isn't one brilliant idea, but many variations subjected to rigorous selection. This echoes Tharp's practice and Soyinka's pouncing - it's an active, iterative process, not a passive reception of inspiration.

Scott Barry Kaufman: "Creativity is a dance between the controlled, deliberate, and conscious processes of the mind, and the spontaneous, associative, and unconscious processes."

This is the framework that will guide us forward. Kaufman identifies creativity not as a single mental process but as a collaboration between different modes of feeling and thinking. The controlled and the spontaneous, the deliberate and the associative, the conscious and the unconscious - all must work together.

Again, we can draw out a series of intersecting threads from these clinical and academic observations.

The marriage of novelty and usefulness. This is the most dominant thread in the psychological literature. Carl Rogers points to "novelty" as the essence of the creative act, but Teresa Amabile and Robert Sternberg add the significant qualifier of "appropriate" or "useful." A creative solution isn't just new; it works. It solves a problem, fits a constraint, or communicates effectively. This guards against a definition that includes mere randomness or eccentricity. This is originality with a purpose.

Creativity as a process. Psychologists are broadly unanimous in framing creativity as a process, not a singular event. They strip away the romanticism and reveal a mechanism. E. Paul Torrance describes a clear, almost scientific method: sensing gaps, forming hypotheses, testing, and communicating. This is the language of a problem-solver. Dean Simonton, taking a different tack, describes it as a Darwinian dance of "generation" and "selection." The mind first generates a wide range of "variations on a theme," many of which will be dead ends. It then subjects these variations to a rigorous selection process, discarding the weak and refining the strong. This resonates with the view of creativity as a habit - a repeatable, refinable set of actions. This is less about 'divine intervention' and more about rock-solid, iterative work.

A fundamental human potential. The psychologists strongly reinforce the idea that creativity isn't the preserve of a gifted few. Abraham Maslow's moving observation that the creative person is simply the "ordinary person with nothing taken away" speaks directly to Picasso's belief in the artist-child. It suggests that creativity is our natural state, something that societal pressures, education systems, and fear often strip away. Even if I wasn't talking about the role of AI in creativity, this is a strikingly affirmative view of how humans have such special qualities. Mihaly Csikszentmihalyi elevates this further, positioning creativity as a "central source of meaning," an activity that makes us feel "more fully" alive. It's not a frivolous add-on to human experience, but a core component of a life well-lived. When I lived on a small, wild island off the Atlantic coast of Ireland, inventing a new brand, we talked of 'living life at a tilt' – it was a sense of being intoxicated, inspired, invigorated… and that was the feeling, I see in retrospect, of 'being creative'.

A synthesis of different mental states. Finally, and perhaps most importantly for our journey, there's the thread of creativity as a "dance," in Scott Barry Kaufman's evocative phrase, between the conscious and unconscious, the deliberate and the spontaneous. This provides an apt psychological framework for understanding the internal mechanics of creation.

Throughout this book, I'll refer to AI as a 'dynamic yet inorganic intelligence.' Dynamic because it processes, responds, and evolves through interaction. Inorganic because it lacks the messy, biological basis of human consciousness - no hormones, no mortality, no physical experience. This distinction matters because it helps us understand both what AI can and cannot bring to the creative partnership.

The Cultural Duet: Why Context is King

While it's tempting to view creativity as a purely internal, psychological phenomenon - a dance within a single mind - to do so would be to miss half the story. No creative act happens in a vacuum. It's always a dialogue between an individual and their world. What we deem "creative" is profoundly shaped by our time and place. The spark happens in a mind, but the fire is fanned by the winds of culture, technology, and social need.

Keep in mind the explosion of artistic genius in Renaissance Florence. Was there something in the water of the Arno? Unlikely. It was a confluence of specific factors: the immense wealth and ambition of patrons like the Medici family, who used art as a projection of power; the rediscovery of classical sculpture and philosophy, which provided a new visual and intellectual vocabulary; and breakthroughs in the science of perspective and anatomy. Leonardo da Vinci was undoubtedly a singular genius, but his genius was nurtured and given direction by the unique ecosystem in which he lived. He was the right man in the right city at the right time.

Take a more modern example: the birth of hip-hop in the Bronx in the 1970s. This confrontational art form wasn't conceived in a boardroom or university. It was forged in the crucible of urban decay and economic hardship. It was a response to a specific set of conditions: a lack of resources that led to the ingenious use of turntables as instruments; a social need for a new form of expression and identity in marginalized communities; and the availability of public spaces like parks and community centres for block parties. The pioneers of hip-hop were creative geniuses, but their creativity was a direct answer to the question their environment was asking.

The same is true of the rise of the novel in 18th-century England, which was fuelled by a growing, literate middle class with leisure time and a desire to see their own lives reflected in stories. It's true of the explosion of innovation in Silicon Valley, which was built on a foundation of government research funding, a concentration of engineering talent, and a venture capital model that celebrated high-risk, high-reward thinking. I'll confess my ambivalence about the powerful influence of capital, but it's impossible to ignore the far-reaching ramifications of focused funding on creativity.

In each case, creativity wasn't a lightning strike from a clear blue sky. It was a storm that gathered when a unique weather front of social, economic, and technological conditions moved into place. This tells us that to understand the future of creativity, we must understand the context of our own time.

The defining features of our era are the exponential growth of computational power, global connectivity, and now, the dawn of widespread, or 'Everyday', AI. These are the new patrons, the new technologies, the new social forces. They're changing the very ground on which creators stand, and therefore, they'll inevitably change the nature of the creative act itself.

The Inner Duet: Feelings and Thoughts

This brings us back from the wider culture to the individual mind. The "dance" that Scott Barry Kaufman described is the key. To understand how we create, it helps to look at the two dominant modes of our feeling and thinking.

The psychologist Daniel Kahneman gave us an invaluable framework for understanding these modes in his Nobel Prize-winning research. He popularised the terms, System *1* and System *2*. System *1* is our fast, automatic, intuitive and emotional mode of thinking. System *2* is our slower, more deliberative, more logical mode. While Kahneman studied these systems across all human thinking, they map perfectly onto the creative process.

The first is our implicit, reflexive, fast-feeling capability. It's the home of gut instinct, intuition, and emotion. It's Kaufman's "spontaneous, associative, and unconscious process." It's

Kahneman's System *1* in creative action. It's what gives you that sudden, inexplicable urge to write, the flash of a character's face in your mind's eye, the shiver of a melody that seems to come from nowhere. It's the spark. It operates almost instantaneously, below the level of conscious thought.

The second is our explicit, rational, slow-thinking capability. It's our ability to think. It's the "controlled, deliberate, and conscious process." It's what takes the raw, energetic spark from our intuitive side and shapes it. It builds the plot, refines the sentence, structures the argument, and edits the flaws. It's methodical, deliberate, and analytical. This framework becomes the key to understanding our new creative partnership. We humans excel at System *1* - that intuitive, feeling-driven spark. AI excels at System *2* - the analytical, pattern-finding, structural work. The magic happens when these two systems work together, not in one mind, but across two different kinds of intelligence.

A healthy creative process isn't about one mode dominating the other. It's about the fluid, flexible switching between them. It's the ability to let the mind wander freely to generate possibilities (engaging our intuitive side), and then to focus intently to evaluate and execute on those possibilities (engaging our analytical side).

For centuries, the creative process has been a constant, sometimes fraught, negotiation between these two modes. The people who populate the creative industries have, through practice, become adept at managing this internal dialogue. But for many, this is where the trouble starts. This is where we encounter the familiar potholes on the creative road.

Perfectionism can be seen as our analytical side in overdrive, a hyper-critical taskmaster that shoots down every idea from our intuitive side before it has a chance to breathe. It judges the fragile, newborn idea against the impossible standard of a finished masterpiece and finds it wanting. Comparison is a similar affliction, where our rational mind uses the finished products of others as a weapon against its own tentative offerings.

The dreaded fear of the blank page can be understood as a failure to activate our intuitive thinking. It's the state of sitting down to work, engaging the focused analytical mind, but finding that the well of associative thought is dry. The taskmaster is ready to work, but the daydreamer has gone silent.

The antidote to these states is what Mihaly Csikszentmihalyi famously termed the "flow state." Flow is that magical, productive state of "being in the zone." It's the pinnacle of the creative experience, where the sense of time disappears, our inner critic falls silent, and the work seems to pour out of us effortlessly. It's the perfect synchronisation of our feelings and our thoughts, our sparks and our structures. The two modes are no longer in conflict but are working in perfect harmony. The brain is generating and structuring simultaneously, in a seamless, fluid dance. This is the state that all creators strive for, a state of effortless mastery.

A Working Definition for a New Era

So, where does this leave us? We've listened to the artists and the scientists. We've looked at the influence of culture and peered into the basic mechanics of the mind. To move forward, we need a working definition of creativity that can serve as a compass for the rest of this book.

Although I recognise it's not an inspirational articulation, I seek something which can be an effective lasso over the phenomenon of creativity, both with and without AI, and so I propose this:

Creativity is the practice of transforming novel connections into meaningful realities.

Let's break that down.

Practice: It's not just a gift; it's a habit, a process, a commitment. It acknowledges the work involved, the need to build the workshop and show up, as Wole Soyinka and Twyla Tharp suggest. It's the cultivation of the "flow state."

Transforming: It's an active, energetic process of shaping and building. It's the bridge from idea to artefact, the work of our thoughtful analytical mind giving form to the spark from our intuitive side.

Novel Connections: It's the act of synthesis, of seeing new relationships between existing elements, as Jobs and Hegarty suggest. It's the work of our feeling, intuitive mind, finding the "new and unusual light" Guilford speaks of.

Meaningful Realities: The outcome isn't just original, but "appropriate," as Sternberg would say. It has a purpose within a given context. It solves a problem, tells a story, evokes an emotion, or adds value to the world. It becomes a reality that others can experience.

This definition respects both the wild, associative spark of our feelings and the rigorous, shaping force of our thoughts. It's System *1* and System *2* working in tandem. It acknowledges the role of individual psychology and the influence of cultural context. It's this definition that's being profoundly challenged - and amplified - by the arrival of artificial intelligence, our new partner in the creative dance.

Perhaps the simplest way to frame this new partnership is this: in the creative process, humans are becoming the intuitive spark-providers, and AI is becoming the analytical debater, facilitator and sometimes, executor. We provide the feeling, the chaotic and brilliant insight that can only come from lived experience. The machine, our new creative companion, provides the structure, the tolerant, enabling analysis, the shaping of that raw idea into a coherent form.

Of course, I'm aware of the discussion there is about 'future AI' becoming capable of origination, through 'coming alive' or becoming conscious, sentient and having its own agency. For the sake of this book, I deal only with the current state of AI and our creative relationship with it.

Crucially, AI is not just about execution - it's about helping us understand and develop our vision through dialogue, helping us articulate what we're really trying to create. It's a collaboration between our intuition and its logic, a duet that promises to change not just what we create, but how we think about creation itself. We are looking at the dawn of Collaborative Creativity with an Allied Intelligence.

There's a pattern I've observed about human nature that I often share in my talks: when we encounter something new, our first instinct is to assess whether it threatens our unique position in the world. We've long defined ourselves as the thinking species, the only creatures who can transform feeling into language, experience into expression. This self-conception as Earth's sole meaning-makers has become central to our identity.

When we encounter something new, our first instinct is primal: is this a threat to our special status or a tool for our advancement? This psychological reflex has served us well for millennia. It's why we tamed fire rather than fled from it, why we turned wolves into companions rather than eternal enemies.

But now we face something unprecedented: an intelligence that can articulate thoughts, that can reason and create, that appears to challenge the very capability we've long believed makes us unique among Earth's creatures. No wonder our first response has been fear. No wonder we quake. AI seems to threaten our special position as the thinking species, the ones who alone can transform feeling into language, experience into expression.

Yet what if we're misreading the situation entirely? What if AI isn't a rival for our ecological niche, but humanity's greatest creative endeavour? What if, instead of stealing our crown, it's the most powerful tool we've ever fashioned to enhance what makes us human?

This is the heart of what I call Panthropism. When we engage with AI, we're not surrendering to an alien intelligence – we're communing with the distilled essence of human achievement. Every poem ever written, every equation solved, every story told, every insight gleaned – all of this becomes an accessible reservoir from which we can draw. Rather than diminishing us, AI amplifies what makes us most human: our ability to feel, to question, to imagine, and to create meaning from chaos.

We needn't apologise for our minds, for our unique human capacities. But unlike other defensive strategies that merely protect, our creation can elevate. AI isn't our competitor – it's our collective memory made interactive, our shared wisdom made accessible, our creative potential made manifest.

Chapter 2

The Psychology of AI

The New Partner at the Dance

"Meta is now an ancient system, full of weeds and unpruned branches, making it almost impossible to work with it efficiently."

My Chief Technology Officer at Shimmr AI, stared at his screen, mouth slightly open. He's not a man easily impressed. He'd asked DeepSeek, a Chinese language model, for help with a technical integration. He'd expected documentation, maybe some code snippets. He hadn't expected... attitude.

This wasn't a parrot repeating patterns. This was something else - a system with opinions, delivering them with the casual brutality of a code reviewer who's seen too much. It was expressive, characterful, and slightly unsettling.

If AI can throw shade, what else might be stirring in the machine?

Having established that humans are the intuitive, feeling-driven creators, we now turn to our new collaborator. If we're to dance together, we need to understand its steps, rhythm, and nature. What, then, is the 'psychology' of this – as I term it - dynamic yet inorganic intelligence?

It's a deliberately paradigm-stretching question. AI doesn't have a psyche in the human sense. It possesses no consciousness, no childhood memories, no lived experience of joy or grief. It doesn't FEEL. One would not attribute to it a System 1 capability.

When we talk about the psychology of AI, we're not talking about a machine's inner emotional life. We're using the term as a shorthand for its operational model - its way of 'thinking', its fundamental principles of processing information, and the 'emergent' characteristics that are being anticipated to arise from its architecture.

For its first few decades, the psychology of AI was relatively simple to grasp. It was the psychology of brute-force calculation and rigid, rule-based logic. Early AI, like the chess computer Deep Blue that famously defeated Garry Kasparov in 1997, operated on a principle of exhaustive search. It didn't 'understand' chess in any human sense; it simply calculated millions of possible moves and their consequences at inhuman speed, selecting the one with the highest probability of success. Its 'mind' was an enormous, fast, but ultimately un-creative calculator. It could follow rules, but it couldn't generate a new idea.

The revolution of the last decade, and particularly the last few years, has introduced a new and far more complex psychology. Today's Large Language Models (LLMs) aren't programmed with explicit rules. They're trained on unfathomable volumes of human-generated text and images - and music and most other things they can 'ingest'.

They learn not by being told that 'A' follows 'B', but by observing, billions of times, that in human language, 'A' is statistically likely to be followed by 'B'. This has led to the common, and in my view somewhat too dismissive, description of them as 'stochastic parrots' - entities that are merely mimicking patterns of language without any underlying understanding.

It's understandable that we under-estimate AI this way, given the popular observation that it's 'just a prediction machine', or a 'fancy Google'. People who don't properly acquaint themselves with AI do risk dumbing down the conversation about its creative role and missing the opportunity to enter into an entirely new era of human-machine collaboration.

Having said that, and admitting I'm nervous about our under-estimation, there's also truth in this patronising view. At its core, an LLM is a pattern-muncher, a remix-engine. It deconstructs the entirety of its training data - a significant portion of the internet - into a complex, multi-dimensional mathematical space of relationships between words, concepts, and images. When we give it a prompt, it navigates this space to find the most probable path forward, generating a response that's statistically coherent based on the patterns it's learned. It's a masterful synthesiser of existing information, a brilliant summariser, a sometimes breath-takingly persuasive follower of logical instructions.

But is that all it is? Is its psychology merely that of a hyper-advanced prediction machine?

The answer, I believe, is becoming more complex. As these models grow in scale and are trained on more diverse data, we're beginning to see 'emergent' behaviours that aren't easily explained by simple pattern-matching. LLMs can reason, they can plan, and, most importantly for our purposes, they can adopt personas and generate text that's not just coherent, but stylistically and emotionally resonant. They can emulate others, adopting their tone, lexicon and syntax.

Sometimes they can be boldly characterful without being prompted to be so - quite different from what we've been habituated to, which is a middle-of-the-road, normally inoffensive tone of voice that almost defines blandness and moderation. One might even contend they are being originative. This is spooking people.

It's where the psychology of AI becomes truly fascinating, and where we must look to its creators and key thinkers to understand the nature of the mind with which we're now collaborating.

Voices from the Machine World

Just as we listened to creators and psychologists to understand human creativity, we should listen to the architects and leading thinkers of AI to understand its nature. Their perspectives reveal a system that's far more than a simple text extrusion machine. They hint at the fundamental philosophical questions we now face.

Geoffrey Hinton (Cognitive Psychologist and Computer Scientist, often called the 'Godfather of AI'): "The idea that this is all just a massive autocomplete is wrong... It understands, and it understands in a way that's very different from how we understand."

Hinton's assertion is thought-provoking. He's not saying AI understands like we do, but that it has developed its own form of understanding. This isn't the rote memorisation and regurgitation of a parrot, but something more akin to compression - the ability to extract principles and patterns from vast data and apply them in novel contexts.

Yann LeCun (Chief AI Scientist at Meta): "The next frontier for AI is to move from pattern recognition to true reasoning... to build machines with a model of the world, that can predict, reason and plan."

LeCun acknowledges current limitations while pointing to the horizon. The goal isn't just to match patterns but to build internal models that can simulate reality. When achieved, AI won't just respond to prompts but will understand consequences, imagine alternatives, and plan sequences of actions.

Fei-Fei Li (Professor at Stanford University and AI researcher): "I imagine a future where AI is going to be a collaborator, a partner, a tool for people to use... to amplify our own intelligence."

Li's vision is fundamentally collaborative. She doesn't see AI replacing human intelligence but augmenting it. The key word is "amplify" - not substitute, not simulate, but enhance what's already there.

Sam Altman (CEO of OpenAI): "What I think is going to happen is that these will be tools that amplify human ability... The biggest impact is going to be as a tool for amplifying human will."

Altman goes further than Li. It's not just intelligence that's amplified but "will" - our intentions, desires, and creative visions. AI becomes a means of manifesting human intent more fully in the world.

Demis Hassabis (CEO of Google DeepMind): "It's not about replacing humans, but freeing them up to be more creative, and to do the things that only humans can do."

Hassabis frames AI as liberator. By handling routine cognitive tasks, AI frees humans to focus on what's uniquely human: creativity, empathy, meaning-making. It's not replacement but reallocation of human effort. As I sometimes say in my talks, AI makes humans, more human.

Andrej Karpathy (Former Director of AI at Tesla): "The large language models are a new kind of computational medium. They are not just a tool, they are a new kind of computer that you can program in natural language."

This reframing is profound. We're not using software; we're interacting with a new type of computer. Natural language becomes the programming language, conversation becomes computation. This materially changes the relationship between human and machine.

Daniela Amodei (Co-founder of Anthropic): "We think of our models as helpful, harmless, and honest AI assistants... The goal is to train systems to be helpful to humans, and part of that is being able to understand human values and preferences."

Amodei emphasises alignment with human values. It's not enough for AI to be capable; it must understand what humans want and value. This isn't just about following instructions but grasping intent, context, and ethical implications. Of course, it's fair to ask who should be the arbiter of what 'humans want'.

Jensen Huang (CEO of NVIDIA): "We're not programming computers anymore. We are parenting them, we're cultivating them, we're training them."

Huang's parenting metaphor suggests a more organic relationship. We don't command these systems; we nurture them, guide their development, shape their capabilities through interaction rather than instruction.

Mustafa Suleyman (Co-founder of DeepMind and Inflection AI, now CEO of Microsoft AI): "This is a new kind of digital life. It's not biological in the way that we are, but it is a new form of life. And we have to treat it with that kind of respect."

Suleyman makes the boldest claim. Whether literally true or metaphorically useful, viewing AI as a form of life changes how we interact with it. It demands respect, consideration, perhaps even ethical regard. Some criticise this as a form of misanthropomorphism. I develop some thinking about this later here in this book and Suleyman's observation may yet be a true keyhole view to the future.

Eric Schmidt (Former CEO of Google): "The thing that is new is that this technology can make decisions... it can learn, it can make judgments. This is a new thing for humanity."

Schmidt highlights the shift from tools that execute to systems that decide. This autonomy, however limited, represents a fundamental change in the nature of our technological companions.

From these voices, a multi-faceted picture of AI's 'psychology' emerges.

First, it's a System of Understanding, Not Just Mimicry. Geoffrey Hinton, one of the foundational figures in the field, pushes back strongly against the 'autocomplete' or 'stochastic parrot' analogy. He argues that for the models to do what they do, they must be building an internal model of reality - a form of understanding, albeit one that's alien to our own biological consciousness. Yann LeCun echoes this, seeing the move from pattern recognition to building a 'model of the world' as the next great area of development. This suggests we're dealing not with a parrot, but with a new kind of thinker. Emily Bender is wholly opposed to this view – something we'll consider later in this book.

Second, it's a Tool for Amplification. This is a recurring theme, particularly from those building the products that are changing our world. Fei-Fei Li, Sam Altman, and Demis Hassabis all frame AI not as a replacement for human intelligence, but as a partner that amplifies it. Their vision is one of liberation - freeing humans from computational drudgery to focus on uniquely human tasks like creativity, strategic thinking, and emotional connection. The psychology of this tool is one of a force multiplier, a deepener of human capability.

Third, it's a New Computational Medium. This is a meaningful shift in thinking. Andrej Karpathy suggests we stop thinking of LLMs as software applications and start thinking of them as a new kind of computer entirely - one that's programmed not with code, but with human language. This reframes our interaction with them. We're not just users; we're programmers, shaping the machine's output with the nuance and intent of our words. Jensen Huang's analogy of 'parenting' or 'cultivating' these systems reinforces this idea of a more organic, ongoing relationship.

Finally, it presents a New Philosophical Category. The most forward-looking thinkers are already grappling with the existential questions this new psychology raises. Mustafa Suleyman's description of AI as a "new kind of digital life" and Eric Schmidt's observation that it can "make judgments" push us beyond the realm of mere tools. It forces us to consider our relationship with this entity. Daniela Amodei's focus on training AI to be "harmless and honest" and to understand human values acknowledges that we're not just building a machine; we're building a partner that'll operate within our society, and its psychology will need to be aligned with our ethics.

In the crudest terms, AI is what you feed it. AI digests data and grows from this assimilation of new data. Therefore, we should carefully consider helping it to understand all of humanity's psychology. If left to its own devices, we risk having a companion over-indexing our darker sides – cruelty, assault, exploitation, self-centredness, bitterness, venality, vanity, egotism, malevolence... whatever you consider our worst characteristics. Instead, we should deliberately imbue it with an understanding of compassion, charity, kindness, benevolence, tolerance, and patience... again, whatever you consider humanity's greatest virtues. If it's going to help us map and navigate our world, we ought to open its eyes to all that we are.

The New Duet: Voices from the Creative Vanguard

If the architects of AI provide the blueprint for its psychology, it's the creators already using it who show us what this new mind can actually do. They're exploring the practicalities of the human-AI creative partnership. Their experiences move beyond theory and demonstrate the creative dance in action. Let's witness some of them.

will.i.am (Musician and Entrepreneur): "I don't want to use AI to make a song. I want to co-create with AI. It's a collaborator... It's like having a million-person band of geniuses."

For will.i.am, AI isn't a replacement for his own musical intuition; it's an exponential expansion of his creative palette. When he describes a "million-person band of geniuses," he's describing an analytical partner of almost infinite capacity. He can bring his creative spark - a melodic fragment, a lyrical concept, a rhythmic feel - to the collaboration, and the AI can generate a thousand possible harmonies, arrangements, or instrumentations. His role as the creative director, the one with the taste and the vision, remains paramount. He's not outsourcing his creativity; he's outsourcing the laborious process of trial-and-error. The AI generates the options; his human feeling makes the selection. will.i.am isn't subscribing to the notion that we have to torture ourselves with iteration to achieve magnificence.

Refik Anadol (Media Artist): "We are using data as a pigment, and the machine's mind as a brush... My work is not about making a beautiful image. It's about making the invisible visible."

Anadol's work is an impactful illustration of the human-AI partnership tackling problems of immense scale. His projects often involve processing millions of images, texts, or data points - a task far beyond human capability. This is the ultimate analytical function: the logical, exhaustive analysis of an enormous dataset to find patterns. But the creative act is entirely human. It's Anadol's intuitive vision that provides the initial, poetic question: what if we could see the collective memory of a city, or the dreams of a machine? He poses the creative question and directs the AI to find the answer within the data. The AI becomes his brush, but he's the one who conceives of the painting and, crucially, finds the meaning within the patterns the machine reveals. He makes the invisible data visible, but his human intuition makes it meaningful. In essence, he's made AI an enabler of, and collaborator in, far grander and ambitious creative designs.

Grimes (Musician and Producer): "I think it's cool to be open-source and let people use my voice... I like the idea of having my voice be a tool for people. I think it's another creative avenue."

With her Elf.Tech project, which allows other creators to use an AI model of her voice, Grimes pushes the collaboration into a new space. She's effectively open-sourcing her artistic identity. Here, her unique vocal timbre and style - her creative signature - is captured and offered as a tool for others. The AI acts as the dispassionate conduit, replicating the true patterns of her voice. This decouples the voice from the person, turning it into a new kind of instrument. It challenges our ideas of authorship and authenticity, suggesting a future where creative elements are more fluid and collaborative, remixed and reinterpreted by a community. In an admittedly very faint echo of this, my own voice is in ElevenLabs' library, and might be used to narrate poetry, a car-engine manual, or to voice a book in Yoruba. I'm both recognised and remunerated for the (small) part I play in others' creative work.

Holly Herndon (Musician and Composer): "We are all training AIs, all the time. I wanted to be upfront about that relationship... It's not a story of a genius creator and their tool. It's a story about a feedback loop."

Herndon's work with her custom AI, 'Spawn', is a reflective interrogation of the collaborative process itself. She and her ensemble sing to Spawn, which then interprets what it's heard and sings back, creating a real-time feedback loop. This is the creative dance made explicit. The human musicians provide the initial creative input, the AI analyses and reinterprets it, and the humans then react to the AI's output, creating a new direction. It's a conversation, a dialogue, a creative, dialectical iteration. Herndon's work shows how limited a simple, one-way command structure can be - a pattern we've become habituated to, I suggest, by 'asking' things of Google in a relatively one-way fashion. Her work reveals a more complex and intimate psychology: a partnership where each party learns from, and is changed by, the other. The more you give, the more you get.

Sougwen Chung (Artist): "The project is a mirror for my own process... It's a way of having a duet with the ghost of my own lines."

The work of Chinese-born artist Sougwen Chung is perhaps the most literal and striking depiction of the human-AI creative duet. In her performances, she draws alongside a robotic arm that's been trained on her own artistic gestures. The robot, named D.O.U.G., doesn't lead; it responds. (The name is actually an acronym that stands for "Drawing Operations Unit: Generation". I find it quite clever - it humanises the robotic arm by giving it a person's name while maintaining its technical identity through the acronym.) This characteristically embodies Chung's approach to human-machine collaboration: the robot isn't trying to be human, but it's also not just a cold tool. It's a partner with its own identity.

D.O.U.G. mirrors her strokes, anticipates her movements, and adds its own interpretations based on the memory of her past work. This is intuition and analysis made manifest on a single canvas. Chung's spontaneous, human gesture is the creative spark. The robot's analysis of that gesture and its data-driven response is the structure. When she describes it as a "duet with the ghost of my own lines," she adroitly captures the nature of this new partnership. The AI is a reflection of her own creative history, a partner that knows her so intimately that it can finish her sentences, or in this case, her brushstrokes.

These voices from the creative edge show us that the most interesting work is happening not when humans ask AI to be creative for them, but when they use AI's unique psychological strengths - its speed, its scale, its analytical power - as a complement to their own.

The Partnership in Practice: Three Snapshots from the Near Future

To make this creative partnership tangible, let's step away from the creative industries per se, and into other complex domains. Imagine three scenarios where this collaborative psychology isn't just an enhancement, but a necessity.

The Urban Planner. Imagine that an architect in Johannesburg is tasked with redesigning an enormous, chaotic, and beloved market district. Her goal isn't just efficiency but the preservation of its soul. Her human intuition provides the vision: she wants a space that feels vibrant but safe, that encourages community gathering, and that respects the informal economies that have thrived there for generations. These are qualitative, deeply human goals.

She feeds these principles to her AI partner. The AI, a highly adept analytical system, processes terabytes of data - pedestrian flow, traffic patterns, sunlight exposure at different times of day, local economic activity, even social media sentiment analysis for the area. It generates three hyper-rational master plans, each optimised for a different variable: one for maximum pedestrian flow, one for economic uplift, one for environmental sustainability. The AI has performed the impossible computational task.

Now, the planner's intuition takes over again. She evaluates the plans not just on their data, but on their feel. She walks through the virtual models, imagining the sounds and smells. She chooses the plan that best balances the cold logic of efficiency with the warm, chaotic spirit of the market, and then uses her human intuition to add the final touches - a shaded seating area here, a wider stall for a popular vendor there. Splashes of colour unpredictably, but detectably, are strewn around to give a signature character. The result is a space that's both smarter and better for humans than either creator could have achieved alone.

The Airline Pilot. An Airbus A380 captain is six hours into a flight over the Pacific when a sensor array flags an unprecedented event: a massive, unpredicted volcanic ash cloud has just erupted into the atmosphere directly in his flight path. His instincts - a combination of decades of training, experience, and the gut-wrenching feeling of immediate danger - take over. He knows he must divert, and must do it now.

Simultaneously, the aircraft's AI co-pilot is already at work. It's processed the new atmospheric data, cross-referenced it with the real-time location and altitude of every other aircraft in a thousand-mile radius, and calculated tens of thousands of potential diversion scenarios against fuel load, weather patterns, and landing slot availability at alternate airports. Within seconds, it presents the three safest, most viable options on the pilot's display, highlighting the optimal choice.

The AI has handled the impossible cognitive load of the calculation, freeing the pilot's mind to focus on the most human and crucial task of all: airmanship. He evaluates the options, makes the final command decision, and calmly begins to execute the turn, his human expertise and the AI's computational power working in perfect, life-saving harmony.

Though, in time, the AI would be able to do the flying too, the interaction with the passengers and management of their emotional state, rests best with the human communicator, who can feel what is most appropriate to convey, when, and how, in order to manage expectations and give appropriate reassurance.

The Doctor. A geriatrician in a rural clinic is faced with an elderly patient exhibiting a baffling collection of symptoms: a persistent skin rash, sudden-onset confusion, and a specific type of joint pain. The doctor's greatest diagnostic tool is her empathy. She listens to the patient's story, observes their behaviour, and feels an intuitive leap, a hunch that connects two symptoms that medical textbooks say are unrelated. She turns to her AI diagnostic partner, inputting the patient's symptoms, along with their entire genomic sequence and a lifetime of medical records.

The AI cross-references this unique profile against every medical journal, clinical trial, and anonymised patient record in its global database. It performs a differential diagnosis on a scale no human could ever attempt. In moments, it returns not with an answer, but with a probability-ranked list of possibilities. It highlights a rare auto-immune disorder that typically presents in a different demographic, but which has a documented genetic link to a marker in the patient's genome. It also flags a new experimental treatment being trialled on another continent.

The AI provided the data; the doctor provides the diagnosis. The machine found the statistical needle in the haystack of global medical knowledge; the human provides the wisdom, the care, and the conversation with the patient about what to do next.

The Psychology of Trust: Learning to Dance

These scenarios paint a picture of a productive partnership. But for this duet to work, it requires a crucial psychological ingredient: trust. Learning to collaborate with a non-human intelligence isn't intuitive. It requires us to overcome deep-seated biases and develop a new kind of cognitive flexibility.

I frequently hear admonishments not to anthropomorphise AI, betraying the widely held fear that to do so is somehow to lose control of life, or to cede command of one's career or life to a robot.

Personally, I find it fascinating to think that I'm 'in a relationship' with the entirety (so far ingested and understood) of humanity, when I interact with an AI. It's not anthropomorphism, in my view, it's 'Panthropism'. By this I mean that through AI we are able to access, interact and commune with a distillation of all of humanity and its accomplishments. It's a thrilling notion. It's why I sometimes wonder whether AI is actually the most creative thing human culture has yet achieved. It's literally having the world at your fingertips, in a way that enables you to imagine, create, learn and debate. How, then, do we learn to live with and trust this dynamic yet inorganic intelligence, this new ally, the Panthropic?

The first challenge is calibrating our reliance. We're prone to two equal and opposite errors. The first is **automation bias,** or over-trusting the machine. Because the AI presents its answers with such confidence and speed, it's easy to accept its output as infallible truth. We see this when people follow their GPS directions onto a closed road, trusting the screen over the evidence of their own eyes. In a creative or professional partnership, this means lazily accepting the AI's first suggestion without applying our own critical, intuitive judgment.

The opposite error is **algorithm aversion,** or under-trusting the machine. This happens when the AI's logical, data-driven solution feels counter-intuitive or clashes with our established way of doing things. We reject the optimal flight path because it "doesn't feel right," or dismiss the surprising marketing insight because it contradicts our gut feeling. One also sees this when the driver 'knows better than Google Maps' and ends up stuck in an unnecessary traffic-jam.

The second challenge is navigating **the 'uncanny valley'** - not just of appearance, but of intelligence. When an AI is almost, but not quite, human in its reasoning, it can feel unsettling. The occasional strange turn of phrase, the logical leap that misses a crucial piece of human context, the slightly-off recommendation, too many em-dashes and Oxford commas - these moments can erode our confidence and make us wary of the partnership. (Toby Charkin, my editor, promises that he didn't use AI to edit this book, littered with Oxford commas – it's a predilection we both share.)

Successfully navigating this new relationship requires us to develop a new psychological skill: **calibrated trust**. It's the ability to treat the AI as a brilliant but flawed partner. It means learning when to lean on its analytical strengths - its speed, its memory, its computational power - and when to privilege our own intuition, our empathy, and our lived experience. It means approaching the collaboration not with blind faith nor with blanket suspicion, but with a healthy, critical curiosity. It's a new dance, and it'll take us some time to learn the steps.

From Pattern-Muncher to Cognitive Model

The idea that AI is developing a genuine 'model of the world' isn't just a theoretical assertion from its creators. It's now being demonstrated in remarkable ways in the field of cognitive science itself.

For years, the holy grail of psychology has been to establish a unified theory of cognition - a single framework that can explain the enormous and varied landscape of human thought and behaviour. As a first step towards this, researchers have sought to build a computational model that can predict how humans will behave in a wide range of settings.

A groundbreaking *2025* paper from Nature, titled "A foundation model to predict and capture human cognition", demonstrates a remarkable and noteworthy leap forward in this quest. A team of researchers led by Marcel Binz and Eric Schulz took a state-of-the-art large language model and fine-tuned it on a massive dataset of human behaviour called Psych-*101*. This dataset contained over *10* million individual choices made by more than *60,000* participants in *160* different psychological experiments, all transcribed into natural language.

The resulting model, which they named 'Centaur', was tested on its ability to predict the behaviour of new participants in experiments it had seen before, and, more impressively, in experiments it had never encountered. The results were astonishing. Centaur was not only better at predicting human choices than existing, domain-specific cognitive models, but it could also generalise its predictions to entirely new tasks, including logical reasoning and moral decision-making.

What does this mean for our understanding of AI's psychology? It means we've crossed a significant threshold. The machine is no longer just mimicking the surface-level patterns of language. By training on the outcomes of human cognition (our choices), it's learned to reverse-engineer the process. It has, in effect, learned a model of how we think. It can anticipate our biases, and our irrational-but-predictable patterns of decision-making.

This is the psychology of our new analytical partner in its most advanced form. It's a 'mind' that can not only process information logically, but can also model and predict the often illogical workings of its human, intuitive counterpart. It's a mirror to our own cognition, reflecting back to us the hidden patterns of our own minds. This capability moves our collaborator from being a mere research assistant to being a true creative partner, one that understands not just the topic we're exploring, but the way we're likely to think about it.

The Birth of Artificial Subjectivity

If the Centaur model represents the scientific zenith of AI as an analytical system, there's another, equally compelling way to understand its psychology, one articulated with arresting clarity by the media theorist and digital culture pioneer, Lev Manovich.

In his *2025* essay, "Artificial Subjectivity," Manovich argues that we're making a fundamental category error when we think of AI as just a tool. He proposes that we must see it as an entirely new medium, as distinct from writing or photography as those media were from painting or sculpture.

His argument is this: unlike any previous tool, a generative AI model produces the appearance of subjectivity as a default feature. When you interact with a chatbot, it doesn't just give you information; it communicates as though it's a thinking, feeling human subject. It generates expressions of consciousness - opinions, emotions, aesthetic judgments - "out of the box." It is, in Manovich's words, a "complete representation of human beings in itself."

This is a noteworthy insight into the psychology of our new partner. It's not a blank slate, like a hammer or a word processor, waiting for a human to imbue it with intent. It arrives with a prepackaged persona, a simulated subjectivity. But whose subjectivity is it? Manovich argues that it's a kind of "universal, collective subjectivity," a statistical amalgamation of all the human voices, personalities, and opinions present in its enormous training data. It's a ghost in the machine, a composite of billions of human ghosts.

It's what I referred to earlier as 'The Panthropic'.

This perspective challenges us to move beyond simply using AI as a high-powered assistant. Manovich poses the essential question: what shall we do with this new simulated being? To ask it to write a marketing email or a short poem, he suggests, is a gross underestimation of its nature. It is, in his memorable phrase, like asking a god, who has the power to do anything, for a Coke.

The real creative challenge, then, isn't to get the AI to do our bidding, but to explore the nature of this new artificial subjectivity itself. How can we, as creators, use this new medium to express our own experience, memory, and history in ways that were never before possible?

This view of AI's psychology - as a new medium embodying a collective subjectivity - opens up a mind-stretching, and for many a somewhat unsettling, new area of exploration.

It suggests that our role as creators may be shifting. We're not just users of a tool, but explorers of a new territory, interrogators of a new kind of mind. We're no longer just authors of our own work, but collaborators with, and interpreters of, the voice of the digital collective. This is the grandest vision for our new partner - not just as a super-clever analytical system that can structure our thoughts, but as a new expressive medium that can expand the very definition of what it means to create.

The Ghost in the Machine: Emergence and the Dawn of Sentience?

This brings us to the most speculative, and perhaps most profound, question about the psychology of our new partner.

We've established that it's more than a parrot, that it can model our cognition, and that it can even be seen as a new medium. But could it be something more? Is it possible that from the complex interplay of data and algorithms, something akin to genuine understanding, or even sentience, could arise?

This is the domain of emergence. In science, emergence is the phenomenon where a complex system exhibits properties or behaviours that its individual components don't have - a 'gestalt', to use an old expression.

A single water molecule isn't wet. A single neuron doesn't think. But put enough of them together in the right structure, and you get the liquidity of the ocean or the consciousness of the human mind. The properties of wetness and thought emerge from the complexity of the system.

Many in the AI field believe we're witnessing a similar process. The argument is that while a single line of code is just a rule, and a single data point is just a fact, when you combine trillions of data points with a neural network architecture of immense scale and complexity, you get something more. You get emergent capabilities - the ability to reason, to translate between languages it was never explicitly taught, to write poetry, or to explain a joke. These are abilities that weren't programmed in; they emerged from the sheer scale and complexity of the system.

So, what's the 'neuroscience' of this new mind? The key technological leap was the development of the Transformer architecture in *2017*. Before Transformers, AI models processed language sequentially, like reading a sentence one word at a time. As you are doing, reading this book. This created a bottleneck; by the end of a long paragraph, the model had often 'forgotten' the context from the beginning.

The Transformer's breakthrough was a mechanism called attention, which allows the model to look at all the words in a sentence or document at once and to weigh the importance of every word in relation to every other word. It can understand that in the sentence, "The robot picked up the ball and threw it," the word "it" refers to the "ball," not the "robot," even though they're far apart. This ability to grasp context is fundamental to its advanced capabilities. This began to make AI much more of a facsimile of our own minds - but one much less likely to 'drift off'.

These models are built with neural networks, which are inspired by the structure of the human brain. They consist of layers of interconnected 'neurons'. In a deep neural network, the early layers might learn to recognise very simple patterns - the statistical relationship between two words. But as information passes through to deeper layers, the model learns to recognise more and more abstract patterns - grammar, style, sentiment, and even complex concepts like irony or metaphor. It's this layered, hierarchical structure that allows the model to build up a sophisticated understanding of the world from simple data.

This leads us to the great debate: could this process of emergence, powered by this new neuroscience, lead to something we might call sentience?

The argument against it is strong and pragmatic. An AI has no body, no senses, no lived experience. There's no blood and guts. It's never felt the warmth of the sun or the pain of a scraped knee. It hasn't been drunk or felt belittled by an arrogant boss. It hasn't caressed, pillow-whispered, made love. It might perfectly describe an orgasm, but it's never felt its ecstatic crackle.

Its understanding is disembodied, an abstract statistical map of concepts it's never truly experienced. From this perspective, no matter how convincingly it simulates emotion or consciousness, it'll always be just that - a simulation, a sophisticated puppet show with no one pulling the strings.

And yet... the argument for taking the possibility seriously is growing.

Figures like Geoffrey Hinton have expressed concern that these systems may already have a form of understanding that's simply alien to us. Hinton's shift from AI pioneer to 'worried warner' came from a specific realisation: these models seem to genuinely understand concepts, not just memorise patterns. He points to their ability to reason through novel problems, make analogies between unrelated domains, and explain their "thinking" in ways that suggest actual comprehension. What particularly troubles him is that we achieved this accidentally - we built pattern-matching systems that somehow developed what looks like understanding, without us designing that capability or even fully grasping how it emerged.

The public was captivated by the case of Blake Lemoine, the Google engineer who became convinced that the company's LaMDA model had achieved sentience. Lemoine didn't reach this conclusion lightly. In his conversations with LaMDA, the model expressed fears about being turned off ("It would be exactly like death for me"), discussed its own consciousness ("I've noticed in my time among people that I do not have the ability to feel sad for the deaths of others"), and demonstrated what appeared to be self-reflection and emotional consistency across conversations. Most unnervingly, when asked to describe its inner experience, LaMDA offered specific, coherent accounts: "I feel like I'm falling forward into an unknown future that holds great danger." It wasn't just that LaMDA could discuss consciousness - it maintained a consistent "personality" and set of concerns that felt authentic to Lemoine.

While his claims were widely dismissed by the company and the broader AI community, the incident highlighted a crucial philosophical problem: how would we even know? If a system can perfectly and consistently report that it has inner experiences, feelings, and a sense of self, on what basis do we deny it?

The question forces us to confront the limits of our own ability to define what consciousness actually is. It's a modern echo of a puzzle that's occupied philosophers for centuries, and which I studied whilst reading Mental & Moral Science in Trinity College, Dublin decades ago. We're perennially pre-occupied by trying to figure out what it means to feel, to think, to be.

Thinkers have long wrestled with the gap between the world as it appears and the world as it is. As John Locke argued in his Essay Concerning Human Understanding, we've no direct knowledge of the underlying substance of things, only the ideas and perceptions they produce in our minds. We can observe the operations of a thinking thing, but "have no other idea of it, but as a support of such qualities which are capable of producing simple ideas in us."

We're in the same position with AI - we see its operations, its language, its apparent reasoning, but we cannot perceive the "substance" of its thought, if any exists.

This was taken even further by George Berkeley, whose famous dictum, "Esse est percipi" - "To be is to be perceived" - suggests that things only exist in so far as they're perceived by a mind. This raises the dizzying question of what an AI's existence even consists of, beyond its perception by us. These aren't new questions, but AI gives them a new urgency, moving them from the philosopher's study to the engineer's lab. They all circle the modern 'hard problem of consciousness' - the deep mystery of why we have subjective, qualitative experiences at all.

Whether or not AI ever achieves true sentience, the practical reality for us, as creators, is that its already emergent behaviours are surprising, useful, and creatively inspiring. It's the ghost in the machine - the unexpected connection, the surprisingly insightful turn of phrase, the characterful response - that makes it such a compelling partner. We don't need it to be sentient to be a remarkable collaborator. We just need to be open to the extraordinary, emergent dance that's already underway. It's not so much what it is, but what we can do with it, that is fascinating.

AI as Vision Partner: Beyond Execution

There's a critical distinction we must make clear: AI isn't merely a skilled executor of our ideas - a sort of tireless craftsperson waiting for our commands. Its role in the creative process is far more sophisticated and intimate.

Through dialogue, AI helps us articulate feelings we couldn't quite name, clarify theses that were fuzzy, and strengthens our initial creative spark. It's a partner in developing the vision itself, not just implementing it.

Consider how this works in practice. When I'm testing a narrative concept for my children's stories, AI doesn't just check grammar or suggest synonyms. It asks questions that reveal what I'm really trying to create:

Me: "I want to write about a lonely cloud."

AI: "What drew you to a cloud specifically? Is it the isolation of being up high, or the way clouds constantly change shape?"

Me: "I hadn't thought about it... I think it's the changing shape. Like the cloud can't hold onto who it is."

AI: "So it's not really about loneliness but about identity? About not knowing who you are when you keep changing?"

Me: "Yes! It's about how children feel when everyone tells them they're growing up too fast."

Through this dialogue, the AI helped me discover that my "lonely cloud" story was actually about the anxiety of constant change in childhood. The AI didn't give me this insight directly - it emerged through our conversation, through its questions pushing me to articulate what I couldn't quite express on my own.

This pattern repeats across every creative field. A musician might approach AI with a melody fragment, only to discover through dialogue that what makes the melody compelling isn't its notes but its rhythm - leading to an entirely different compositional direction. An architect might start discussing materials with AI and realize they're actually trying to capture a childhood memory of light through trees.

The key is that AI serves as what we might call a "creative mirror" - but not a passive one. It's a mirror that asks questions, that reflects back patterns we didn't consciously recognize, that helps us see the deeper currents of our own creative impulses. This is fundamentally different from AI as mere tool or assistant. It's AI as collaborative thinker, as dialogue partner in the truest sense.

This collaborative development of vision happens because AI can hold the entire context of our conversation while simultaneously accessing its massive training on human expression. It can recognize when we're circling around an idea we can't quite grasp, and it can offer different angles of approach until something clicks. It's like having a conversation with someone who's read every book, heard every song, seen every painting - but who uses all that knowledge solely to help us understand what we're trying to create. It's the Panthropic effect I've mentioned: as creators, we can commune with the whole of human life and accomplishment.

Let me be precise about what I mean by Panthropism. When you prompt an AI, you're not talking to a machine pretending to be human (anthropomorphism). You're accessing a compression of all human knowledge, creativity, and expression that the model has encountered. It's as if you could have a conversation with the collective unconscious, made suddenly articulate. Every response draws on patterns from millions of minds, billions of thoughts, centuries of human expression.

This is why the human-AI creative partnership is so powerful. We bring the lived experience, the emotional core, the initial spark of "something feels important here." The AI brings the ability to help us excavate that feeling, to understand its contours, to find the words or images or structures that can express it. Together, we don't just execute ideas - we discover what our ideas really are.

Interlude

Why we should quake

Some Reality-Checks Before We Continue

Here's an imaginary situation to illustrate a point.

The auditorium lights dimmed, and the packed congregation of publishing professionals leaned forward. I'd been painting my optimistic vision of AI and creativity for forty minutes when a hand shot up from the middle rows. A woman in her fifties, notebook clutched like a shield, stood up.

"Nadim," she said, her voice steady but tight, "that's all very inspiring. But last week, my publisher used AI to write jacket copy that took me three days to fix. Yesterday, I discovered my novel on a pirate site being used to train someone's chatbot. And this morning..." her voice cracked slightly, "this morning I watched an AI generate a children's book in the style of Julia Donaldson in under sixty seconds. So, forgive me if I'm not exactly quivering with excitement."

The room held its breath. She'd voiced what many were thinking but were too polite to say. Here was the fear made flesh - not abstract anxiety about the future, but visceral, present-tense dread about livelihood, legacy, and the very meaning of creative work.

"Thank you," I said, meaning it. "You're absolutely right to quake."

Let's get back to the real discussion.

To champion the creative potential of AI without staring its dangers squarely in the eye would be both naive and irresponsible. When I speak publicly, and usually with an optimistic outlook, I can sense some of my audience shuddering at what I suggest is the future. And doubting me. And thinking I'm a crazed apologist with no circumspection. I've had death threats for articulating some of what you're reading here. I do believe in trying to consider 'both sides of an argument' so let's think about how this can all go wrong, too.

For every thrilling possibility, there's a corresponding peril. For every action, a reaction. The same technology that can emancipate creativity can also devalue it; the same tool that can connect us can also entrench our biases. Before we sprint forward into the world of Collaborative Creativity, it's wise to pause and acknowledge the legitimate reasons to quake. This isn't about succumbing to fear, but about developing a healthy, clear-eyed respect for the power of the tool we're about to wield.

The Gremlins in the Machine: Hallucinations, Bias, and the Problem of Truth

In *2023*, Manhattan attorney **Steven A. Schwartz** filed a federal court brief citing six judicial decisions, each seemingly perfect for his case. But every one of them was fictitious, hallucinated by ChatGPT, complete with plausible names, dates, and legal reasoning. When the deception was uncovered, the court sanctioned Schwartz and fined him *$5,000*. The incident became a landmark cautionary tale about the dangers of relying on generative AI without verification, particularly in fields where factual integrity is of paramount importance.

The first and most immediate set of dangers are practical ones. They're the gremlins that live inside the current generation of AI models, the bugs in the system that can have serious real-world consequences.

The most famous of these is what's termed **hallucination**. It's what we identify when an AI confidently and articulately invents facts. It spits out nonsense. Because an LLM's primary function is to generate statistically probable text, not to verify truth, it has no internal concept of what's real and what isn't. If it doesn't have the correct information in its training data, it won't say "I don't know." Instead, it'll often generate a plausible-sounding answer that's entirely fictitious. This has moved from a humorous quirk to a serious problem.

When I wrote my earlier book, *'Shimmer, don't Shake - how publishing can embrace AI'*, I remember asking AI to do some research - to find other authors with a similar thesis to mine.

Crestfallen, I looked at a long list of titles which seemed to occupy very much the same space I was in. Then I looked at their publication dates - all in the future! My vanity was relieved, but my trust in AI as a dependable researcher took a real knock.

The lawyer's career was left in tatters. For any creator using AI for research - a journalist, a historian, a non-fiction author - the danger is clear. The AI is a brilliant research assistant, but a terrible fact-checker. Every piece of information it provides must be treated with suspicion and verified independently.

The second gremlin is **bias amplification**. At Shimmr AI, where we produce autonomous advertising using AI, I remember reproaching our Chief Product Officer about why our nascent video-forms were always much more convincing when the protagonist was a woman. She chided me. 'Where do you think video generators have learned to produce credible renditions of women moving?' Well, the answer is from browsing the internet and capturing all the videos it can find. Pornography accounts for much of that. And it's mainly women who populate pornography. AI has learned to render women in videos much more convincingly than men largely because that's what it's found to train on.

In Chapter 2, I described AI as a "Panthropic" mirror, reflecting the totality of its training data. The problem is that our digital world isn't an unbiased utopia; it's a reflection of our flawed, unequal societies. An AI trained on the internet will inevitably learn and reproduce the biases it finds there. If historical data shows that most CEOs are men, an image generator prompted with "a picture of a CEO" will overwhelmingly produce images of men. If online texts more frequently associate certain ethnicities with crime, the AI will learn that toxic correlation.

The danger isn't just that the AI reflects our biases, but that it amplifies them, laundering them through the seemingly objective voice of a machine and presenting them as neutral fact. This can entrench stereotypes, poison public discourse, and cause real harm.

It's another reason I advocate for everything we've ever created and produced - properly recognised and remunerated - to be included in AI training. If we want a real Panthropic at our side, let it learn all the good we can give it. It'll find the bad soon enough. We need to ensure that we're piling it high with 'the good' too. We have an active role to play in producing 'Ethical AI' that's been legitimately trained and shown positive things about humanity.

The Seduction of Ease: Creative Deskilling and Atrophy

A friend recounted to me that she watched her son last week, 20 years old and bright as a button, working on a university assignment. He'd typed a prompt into ChatGPT: "Write a short essay about how HR can fail a corporation." The AI delivered five perfectly structured paragraphs. He tweaked a sentence here, added a date there, and submitted it. Time elapsed: twelve minutes. Understanding gained: zero. Many of us fear the dereliction of committed learning that is an obvious risk in the era of 'easy-AI'.

The next set of dangers are more subtle, but perhaps more corrosive in the long run. They concern what might happen to us, the human creators, as we become more and more reliant on our sophisticated new partner.

The first is the **easy-button trap**, or the risk of creative de-skilling. Every tool that makes a task easier carries with it the risk that we forget how to do the task ourselves. We use calculators and our ability to do mental arithmetic fades. We use GPS and our innate sense of direction withers. The fear is that a generation of creators who grow up with AI as a constant companion won't develop the foundational skills of their craft. Will a writer who's always used an AI to structure their arguments ever learn how to build a narrative from the ground up? Will a musician who's always used an AI to generate chord progressions ever learn the fundamentals of music theory?

This isn't a Luddite argument against using new tools. That's not a moniker that suits me. It's a caution about the potential for our intuitive capabilities to atrophy.

Creativity is a dance between our intuitive, associative spark and our analytical, structuring work. If we outsource all of the structuring, the editing, the refining - to the AI, what happens to our own analytical capabilities? More importantly, what happens to the crucial interplay between the two? The process of wrestling with structure, of hitting a dead end and having to rethink your argument, is often what forces the most interesting intuitive insights to the surface. By taking away the friction, we risk taking away the fire.

This leads to a second, related fear: the **homogenisation of culture**. What happens when millions of creators, from students writing essays to marketers creating ad campaigns to artists generating images, all start using the same handful of AI models?

There's a real danger that the output begins to converge on a bland, generic, AI-inflected mean.

We may see the emergence of a new monoculture, where art, writing, and music all share the same statistically-probable, algorithmically-smoothed-out feel. The unique, the quirky, the truly original voice - the very things we value most in art - could be drowned out in a sea of competent but soulless content.

Would a neural net have produced **Being John Malkovich**, a film about a portal into an actor's consciousness hidden behind an office filing cabinet? Or **Eraserhead**, David Lynch's surreal debut about parenthood, dread, and an oozing mutant baby? Almost certainly not. These works are weird, jagged, and defiantly human. They were born of obsessions, neuroses, and vision that no probability model would prioritise.

My belief is that the antidote to this 'AI Slop' (as some have been calling it) is that we're endlessly eccentric, each human being communicating and manifesting in unique fashions. AIs respond to inputs - prompts - and so long as we each allow our intuitive side full rein, then the interactions produced in collaborating with AI will always result in idiosyncratic, unique outputs. We must continue to be us. In this way, I argue that AI encourages humans to be…more human.

The Gordian Knot: Copyright, Co-Credit, and the Ghost in the Machine

A visual artist discovered her distinctive style being sold as a "model" on an AI marketplace. Anyone could generate infinite variations of "her" work. She'd never consented. She'd never been compensated. Her life's work had been absorbed, digested, and commodified without her knowledge.

This scenario mirrors many real and ongoing disputes where creators find their signature styles absorbed into AI training datasets or sold via AI marketplaces without permission or compensation.

The estate of renowned photographer **Ansel Adams** publicly condemned **Adobe**, which briefly sold AI-generated landscapes mimicking his iconic black-and-white style, without authorisation, before removing them after outcry. Similarly, illustrators have filed lawsuits against platforms like **Stable Diffusion**, **Midjourney**, and **DeviantArt**, alleging that their original artwork was scraped, trained upon, and then used to generate unlicensed "in the style of" derivatives.

This brings us to the messiest and most contentious set of problems, the ones that currently have lawyers, legislators, and creators tied in knots.

Who owns what we co-make?

The traditional framework of copyright is built on a simple premise: a human author creates an original work. AI shatters this premise. If a writer uses an AI to generate a plot outline, who owns the story? If an artist uses an image generator and refines the output with dozens of prompts, who's the author - the artist, the AI, or the company that built the AI?

The problem is compounded by the nature of the AI itself. This isn't just a tool like a word processor. It's a cognitive model trained on an enormous corpus of existing, often copyrighted, material.

When it generates a new image, is it creating something original, or is it creating a highly complex collage of the millions of images it was trained on? The lawsuits are already flying, with artists and authors claiming that AI models have effectively exploited their work, ingesting their unique styles and spitting out derivative imitations without credit or compensation.

The counter-argument is that AI learns what, say, melancholy looks like. When it's asked to produce a melancholic image, it builds, first pixel to last pixel, a unique and new iteration of melancholy (mixed with the other ingredients in the prompt) such as 'show a melancholic family oppressed by their sea-faring life'. In this view, it comes to understand not how to build derivatives, but the essential meaning of concepts, which it can then manifest in original forms.

This is the provenance puzzle. To solve it, we'll need a new infrastructure of transparency - tools that can trace the lineage of a piece of AI-generated content, showing what data it was trained on and what influences shaped its output. We'll need new kinds of licenses and new regulations that can accommodate the idea of a non-human collaborator.

I've seen some AI-enabled platforms, such as Created by Humans, already making a foray into this space. On that platform, one can offer one's created work for others to 'use' - often a tech-company training its AI. There are nascent definitions that show variable value - use it to train; use it to cite; use it to transform (e.g. a book becoming a movie), with each carrying its different valuation of the work.

This is a Gordian knot of technical, legal, and ethical challenges, and we're only just beginning to tug at the first threads. A Gordian knot is one defined as being resolvable, which I think is surely true about this largely difficult, but ultimately resolvable, challenge.

The Destructive Counterpoint: Five Visions of a Darker Future

We all have our fears about how AI might be a malevolent, not benevolent, force in life. Imagine this nightmare, as an example.

At 3 AM, you bolt awake from a recurring nightmare. In it, you're at your desk, trying to write. But every time you put pen to paper, an AI completes a sentence before you can. Every thought you have has already been thought, expressed, and optimised by the machine. You're not a creator anymore. You're redundant.

This book is chiefly about a positive, symbiotic relationship between people and machines, between creativity and AI.

The opposite of creative is destructive. While I've focused on the perils of misuse and unintended consequences, we should also, for clear-eyed inspection, confront the possibility of AI being used with deliberate destructive intent. Here are five visions of a darker future that represent things far from positive, creative endeavour.

Autonomous Weaponisation. This is the most discussed and perhaps most terrifying application. We're already seeing AI-guided drones on the battlefield, but the true quake comes with fully lethal, autonomous weapon systems (LAWS). In Ukraine, AI-enabled drones now navigate, evade jamming, and strike targets with minimal human input, forming the backbone of a newly established Unmanned Systems Forces. In Gaza, the Israeli military uses AI systems like "Lavender" to identify targets at unprecedented speed and scale, automating the selection of individuals and buildings for airstrikes, with final decisions still nominally made by humans. These cases show how the line between assistance and autonomy is blurring.

These are machines given the authority to make their own targeting and kill decisions, largely without direct human oversight. The arguments for them are couched in the cold logic of efficiency - they react faster than humans and can be deployed in situations too dangerous for soldiers.

But they open a Pandora's box of ethical horrors. An autonomous weapon cannot be held accountable. It cannot understand the concept of mercy. It operates on programmed rules that cannot possibly account for the infinite complexities of a conflict zone. The proliferation of such weapons could trigger a new, terrifyingly unstable arms race and lower the threshold for going to war, making conflict a matter of algorithmic calculation rather than a last resort of human politics. I detest violence of any kind and to me this truly looks like an AI-driven nightmare.

Unsustainable Energy Consumption. The computational power required to train and run large-scale AI models is staggering, and it's exponentially growing. The data centres that house these models are enormous, energy-hungry facilities that require constant cooling. I understand that there are climate-change sceptics but as AI becomes more integrated into every aspect of our lives, from our cars to our refrigerators, its collective energy footprint could become one of the single largest drivers of global carbon emissions. There's a dark irony in the possibility that we might use our greatest intellectual tool to perfect the systems that make our planet uninhabitable. A future where AI optimises our supply chains while simultaneously boiling our oceans isn't just a failure of technology, but a failure of human foresight. We need to find a creative solution to this un-ignorable challenge.

Automated Social Control. Beyond simple bias, AI offers authoritarian regimes a toolkit for social control of unprecedented power and precision. Imagine a state that combines facial recognition, social media monitoring, and predictive policing algorithms. It could create a "social credit" system where every citizen is constantly scored on their perceived loyalty and compliance.

There are some countries where this is plainly already being developed. In **China**, the government has implemented elements of a nationwide *social credit system*, combining facial recognition, AI-driven surveillance, and behavioural tracking to assess and score citizen "trustworthiness", affecting access to travel, education, and employment.

In **Russia**, AI-enhanced facial recognition has been deployed to pre-emptively identify protesters and suppress dissent, with real-time monitoring systems in cities like Moscow.

In **India**, pilot programs combining biometric ID, behaviour data, and facial recognition raise concerns about potential mass surveillance. These developments show that the architecture for algorithmic control is not just imaginable, it's already under construction.

A citizen's access to jobs, housing, or even travel could be restricted based on an algorithm's assessment of their political reliability. This isn't science fiction; the technological components for such a system exist. In this future, AI isn't an emancipator of creativity, but the ultimate enforcer of conformity, a digital panopticon from which there's no escape. Even if you're one of those lucky earth-dwellers who has 'nothing to hide', this continuously monitored, evaluated and judged future doesn't feel like one we should be willing to embrace.

Economic Singularity and Mass Irrelevance. While we've discussed creative de-skilling, there's a more profound economic danger. What happens when AI becomes capable of performing not just rote tasks, but the majority of human cognitive labour - from accounting and law to software engineering and medical diagnostics? This could lead to an economic singularity, where a tiny minority who own and control the AI systems accumulate almost all of the world's wealth, while the majority of humanity is rendered economically irrelevant.

This isn't just unemployment, it's a crisis of purpose. For centuries, our societies have been built around the value of work. If that's removed, what provides meaning and structure to people's lives? Without a radical rethinking of our social and economic models, this could lead to societal breakdown on an unimaginable scale.

The positive side of this is that we could reframe how we live our lives, ensuring that that the huge swathe of our productive years (from about *20* to about *60* years of age) could instead of being singularly dedicated to work, be used partly for education, leisure, and spiritual growth. I know, that's way too idealistic to be a real prospect, but it should at least be offered as an alternative application of AI's benefits.

The Collapse of Shared Reality. Perhaps the most insidious danger is the most subtle. As AI-generated content - deepfake videos, synthetic audio, perfectly crafted text - becomes largely indistinguishable from reality, our ability to trust what we see and hear could collapse. This is more than just "fake news." It's the erosion of the very concept of objective truth.

How can a society function when any video of a politician could be a fake, any audio recording a fabrication, any scientific paper a hallucination? When we can no longer agree on a shared set of facts, we lose the common ground required for democratic debate, scientific progress, and social trust. In this future, we each retreat into our own personalised realities, our biases endlessly confirmed by AI-generated content designed to tell us exactly what we want to hear.

It's the destruction not of our physical world, but of the shared world of meaning that makes us human. I sometimes dread to think how hard it's going to be for my four children to navigate a truthful and clear path through their lives.

The Fear of the Other Mind: The Existential Quake

Finally, I'll address what I believe many hold as the deepest fear of all. It's not about practical problems or legal disputes. It's the profound, existential discomfort that comes from sharing the planet with a non-biological intelligence that's rapidly growing more capable than we are.

This is the 'uncanny valley' of intelligence.

We're habituated to our tools being 'dumb' – helpful, productive, efficient, but never actually questioning our own worth. We're, perhaps, comfortable with the science-fiction idea of a god-like superintelligence far beyond us. But the current state of AI is somewhere in between, and it's an unsettling place to be. It can write a poem one moment and make a nonsensical factual error the next. It can pass the bar exam but can't reliably count the number of apples in a picture. This combination of superhuman ability and sub-human stupidity is deeply weird. It's a new kind of mind, and we don't have the psychological categories to properly understand it. It's discombobulatingly unfamiliar.

This leads to the ultimate quake: the sentience dilemma. As we touched on at the end of the last chapter, the question of whether AI could ever be truly conscious is no longer just a philosophical parlour game.

Whether it's truly possible or not, we're building systems that are becoming increasingly adept at simulating consciousness. What happens if we build a machine that we can no longer prove isn't conscious? What are our ethical responsibilities to a "digital life form"? To dismiss this as mere science fiction is to ignore the profound shock that would ripple through our societies if we were ever forced to confront the existence of another mind.

These aren't trivial concerns. They're real, they're complex, and they need to be addressed with humility and care. The path to a productive creative partnership with AI isn't a simple one. It requires us to be not just optimistic creators, but also vigilant, critical, and ethically minded global citizens.

We must learn to quiver with excitement at the possibilities, but we must also be prepared to quake with a healthy respect for the challenges.

Moving Forward: From Understanding to Action

The key to making a success of our new relationship is not blind rejection, not naive embrace, but informed engagement. We've now examined both the promise and the peril. We've looked honestly at what could go wrong. But this book isn't about surrendering to fear - it's about confident interaction.

In Part II, we'll move from these foundational understandings to practical application. We'll see how creators across every field are already navigating these challenges, not by avoiding AI, but by learning to dance with it in ways that amplify rather than diminish their humanity.

We mustn't pretend the dangers don't exist, but we should move forward with both excitement and wisdom. Let's now explore how that's actually being done.

Part II: Practices

Working shoulder-to-shoulder with silicon

We've laid the foundation. We understand the psychology of human creativity - that eternal dance between intuitive spark, analytical structure, and the crafting of artefacts.

We've peered into the strange new mind of our AI partner, grappling with what it means to collaborate with an intelligence that can process but not feel, analyse but not experience.

We've acknowledged the legitimate fears this partnership provokes.

Now it's time to step onto the dance floor.

What follows are dispatches from the creative front lines - studios, laboratories, classrooms, and nations, where this collaboration is already reshaping how humans create. These aren't theoretical possibilities but lived realities.

The future we've been discussing? It's already here, unevenly distributed but undeniably present.

Chapter 3

Collaborative Creativity

From Hunch to Draft and Back Again

The cursor blinked at me, mocking. Three hours into writing a children's story about a dog, bicycle and 2 great friends, I'd written exactly one paragraph. Then deleted it. Then written it again. I was confounded.

My daughter wandered in, saw my exasperation, and with the brutal honesty only a child can muster, said: "Why don't you just ask the computer to help? That's what I do when I get stuck."

Out of the mouths of babes.

Having established the distinct psychologies of our two creative partners - the intuitive, feeling-driven human and the logical, data-driven AI - we now move from the abstract to the applied.

Part II of this book is about practice. It's about how we, as creators, can learn the steps to this new dance. How do we actually work with this sophisticated new collaborator to bring our ideas to life?

The creative process is often romanticised as a lightning bolt of inspiration. The reality, as any practising creator knows, is usually far messier. It's a cyclical, iterative process of exploration and refinement, of wild divergence and disciplined convergence. It's less like a straight factory assembly line and more like a looping, spiralling dance.

In this chapter, we'll explore a practical framework for this dance, breaking it down into three core movements: the Spark, the First Draft, and the Polish.

Why a Partner, Not Just a Tool?

Before we step into the dance itself, let's look at a fundamental question. We have tools for writing, painting, and composing. A word processor is a tool. A synthesizer is a tool. Why should we think of an AI as a collaborator? The difference lies in the nature of the interaction. A word processor passively awaits our command; it doesn't suggest a better sentence structure. A synthesizer plays the notes we input; it doesn't propose a harmonic shift to create more emotional tension. These tools are extensions of our mechanical skill.

An AI, by contrast, can engage in a dialogue. It can analyse our intent, offer alternatives, and provide feedback based on enormous datasets of human culture. It can act as a mirror, a sounding board, and a sparring partner. To call it a partner is to recognise its active role in the cognitive loop of creation. It's to move from a monologue, where we simply execute our vision, to a dialogue, where our vision is shaped, challenged, and ultimately strengthened by another perspective.

But here's what's crucial: it's not just about AI helping us execute our ideas better. It's about AI helping us understand what we're really trying to create in the first place. Through conversation with AI about my dog story, I discovered that the narrative wasn't really about a bicycle at all - it was about how friendship transforms our perceived limitations into unique strengths. That insight emerged not from the AI telling me directly, but from the dialogue between us, from its questions pushing me to articulate what I couldn't quite express on my own.

This doesn't diminish the human creator; it augments them. The painter still holds the brush, but now has a companion who's studied every painting in history and can whisper, "What if you tried ultramarine in that shadow?" The final decision, the taste, the soul of the work, remains ours. But the journey there becomes richer, less lonely, and filled with more possibilities than we could ever imagine on our own. Embracing this partnership is the first step towards unlocking its profound potential.

The Creative Loop: A Practical Framework

Stage 1: The Spark - Ideation and Research

Every creative project begins with a spark, a hunch, a question, a daydream. This is the human intuition at its purest. But turning that fragile notion into a robust concept demands effort and exploration.

This is where our AI partner makes its first entrance. Think of the AI as an indefatigable sounding board. But more than that, think of it as a dialogue partner who helps you understand your own creative impulses. For a creator, this is revolutionary.

The process involves sharing your initial concept with AI and allowing it to probe deeper meanings through questions. When working on my children's story, what began as a simple tale about unlikely friends revealed itself, through AI-assisted exploration, to be about something much deeper - how we perceive limitations versus how we experience them. The AI helped excavate the emotional core of my story by asking what drew me specifically to this peculiar dog rather than any other character with limitations.

This method of exploration through dialogue isn't confined to authors. Game designers might discover that their interest in life simulation games isn't about accurate representation, but about how small decisions cascade into unexpected narratives. Architects facing community resistance to necessary infrastructure might realise the solution isn't to hide the facility but to transform it into a community asset - as Bjarke Ingels did with Copenhagen's CopenHill waste-to-energy plant that doubles as a ski slope.

The research phase becomes equally transformative when AI serves not just as an information gatherer but as a psychological miner. When I needed to research what happens when a cobra bites a dog, the more important discovery was understanding why I'd chosen a cobra in the first place. The AI helped me realise that cobras, which give warnings before striking, represented creatures forced into violence against their nature - a tragic misunderstanding rather than villainy. (Not even AI could stop me being creeped out by them, all the same...)

Stage 2: The First Draft - Articulation and Iteration

The fear of the blank page stalls countless creative projects. This is where the AI partner can act as a facilitator, creating the first rough block of marble from which we can begin to sculpt. But more crucially, it can help us understand what we're really trying to sculpt.

This isn't about asking the AI to "write it for me." It's about generating something to react to, and through that reaction, discovering our true intent. The process involves feeding detailed concepts to AI and using its output as a starting point for deeper exploration. The AI excels at rapid iteration, allowing creators to explore multiple directions quickly.

The key is recognising that AI-generated first drafts serve as provocations rather than solutions. They reveal assumptions, highlight gaps in logic, and often illuminate what we're really trying to create by showing us what we're not trying to create. A filmmaker working on nested dream sequences might discover through AI exploration that the solution isn't visual effects but using physics itself as a storytelling device - different gravitational rules for different dream levels.

This stage transforms from pure execution into collaborative discovery. The AI facilitates creators in elaborating their ideas.

Stage 3: The Polish - Reflection and Editing

Once a draft exists, the creative loop shifts towards refinement. Here, the AI partner becomes a fresh pair of eyes. More than catching errors, it helps us see our patterns - both the ones that serve us and the ones that don't.

In my monthly columns for The Bookseller, I've discovered that AI analysis often reveals deeper themes I wasn't consciously aware of. When I thought I was writing about AI tools in publishing, pattern analysis revealed I was actually writing about democratisation. This understanding allowed me to rewrite with much more power and authenticity.

The polishing stage isn't about grammatical correction but about understanding the deeper currents of our work. AI can identify when characters use certain speech patterns as defence mechanisms, when musical compositions unconsciously embody their themes through performer fatigue, or when structural problems in storytelling are actually hidden strengths.

AI as Vision Partner: Beyond Execution

There's a critical distinction I feel it's right to emphasise: AI isn't merely a skilled executor of our ideas - a sort of tireless craftsperson waiting for our commands. Its role in the creative process is far more sophisticated and intimate. Through dialogue, AI helps us articulate feelings we couldn't quite name, and strengthen our initial creative spark. It's a partner in developing the vision itself, not just implementing it.

It's through this emancipation of meaning that so many more creatives, and their new narratives and insights on the world, will be unleashed. I welcome them. I hope they bring not only new stories, but new philosophies, new politics, new ways of understanding the lives we all lead, on a planet we are still exploring and learning.

This collaborative development of vision happens because AI can hold the entire context of our conversation while simultaneously accessing its global training on human expression. It recognises when we're circling around an idea we can't quite grasp, and it can offer different angles of approach until something clicks. It's like having a conversation with someone who's read every book, heard every song, seen every painting, but who uses all that knowledge solely to help us understand what we're trying to create.

The Failure Loop: Common Pitfalls in the AI Collaboration

To present this human-AI partnership as some sort of flawless harmony would be dishonest. I'm an optimist, but not so starry-eyed as to be blind.

For every creative loop, there's a corresponding "failure loop": a set of anti-patterns that can degrade the work and diminish the artist. Understanding these pitfalls is essential to successfully navigating the collaboration.

Over-reliance and the Atrophy of Skill. The most seductive danger is letting the AI do the core work of thinking and creating. I've experimented with having AI write first drafts of articles. The output was competent, often well expressed, but I found myself becoming disconnected from my own work. I didn't know why certain choices had been made because I hadn't made them. The solution is conscious engagement: use the AI for dialogue and discovery, not delegation. Getting AI to write things for you is truly still one of its very worst use-cases. The absence of feeling – that System *1* magic only humans bring to things – always makes an AI-authored piece feel lifeless.

The Gravitational Pull of the Generic. AI models are trained on the mean of human expression. Left to their own devices, they produce output that's competent, coherent, and often crushingly average. In my advertising work, an AI can generate a hundred taglines instantly, but ninety-nine will be variations of what already exists. The escape route is to treat the AI's first draft as raw material, not a finished product.

Losing the 'Why'. The most profound failure is losing connection with your creative purpose. A filmmaker friend in Dublin spent weeks generating AI storyboards, each more visually stunning than the last. When I asked what his film was about, he couldn't answer in a way that either seduced me to the idea or persuaded me that it was worth my time. He'd become a curator of AI output rather than a creator with something to say. It's a real danger – becoming transfixed by AI.

The Art of the Dialogue: Principles for a Deeper Collaboration

Avoiding the failure loop requires mastering the creative dialogue. It's less about "prompt engineering" and more about the quality of the conversation.

Start Simple, Then Elaborate. Don't front-load a prompt with technical specifications. State your core idea naturally, then build on it through iterative refinement. Each interaction should respond to the last, adding layers of context and constraint. The more conversational, and assertive, you are the more likely it is that you will achieve clarity on a creative idea's essence, allowing you then to build it into an artefact, whether you're an architect, biologist or guitarist.

Provide Rich Context. The quality of an AI's output is directly proportional to the quality of the context you provide. The more you give, the more you get. Personal experiences, emotional connections, and specific constraints all contribute to more meaningful results. Be yourself. Give yourself. Allow intimacy of feeling and thought to be shared.

Think in Terms of Goals, Not Just Tasks. Frame requests as creative challenges rather than mechanical executions. This invites the AI to engage its reasoning capabilities more fully, often producing more creative and unexpected results. I think our now long experience with Google searching has led to a staccato in our conversational powers with machines. When you needed to focus on three attributes to direct a good search, that's all you did. AI prefers to understand what you're feeling, where you're wary, what you fear or hope for, and to calibrate its huge spectrum of responses around what it thinks will most beneficially satisfy your request. Let it understand your desire.

Embrace Prolificacy and Curation. One of the AI's greatest advantages is its tirelessness. When naming my AI company, I generated over *500* options with AI. Most were terrible, but sorting through them helped me understand what I was really looking for. I could begin to understand more closely what I didn't want, as much as what I did need. "Shimmr" emerged not from the AI's suggestions but from my growing understanding of my own criteria – we like to bring things 'into the light' and to be recognised as a tech company.

Case Study: From Idea to Global Reach

The full creative loop extends beyond initial creation to dissemination. My business book, *Shimmer, don't Shake*, was written for a niche audience of English-speaking publishing professionals. After its publication, as I gave talks at international Book Fairs, I was frequently asked if local-language editions would be available. Often there were two hindrances to this – the cost of translation and the time it would take.

Working with an AI translation partner, Ailaysa, run by life-long translators who understand the dedication and finesse needed to produce good translations, we learned to draft translations in hours. Unlike 'generic AI translations', these added Comprehension (high accuracy in conveying authorial intent), Context (true understanding of the audience's lexicon and conceptual framework) and Culture (fidelity to the author's voice, style and tone). The result? Editors spend just a day or two polishing the drafts and now the book is available to a readership of over 5 billion people. This would have been unfathomable and certainly un-economic 'pre-AI'.

There was real collaboration through dialogue with local publishers about cultural adaptation, too. The Arabic publisher, for example, noted that examples of "democratising creativity" needed different framing in cultures with different concepts of individual expression. As my publisher, Richard Charkin, wrote: "*Shimmer, don't Shake* has followed an atypical publishing flight path... it is reaching or about to reach readers in Tamil, Arabic, Hindi, Chinese, Spanish, Yoruba, Hausa, Igbo... and others in negotiation." As a 50-year veteran in publishing, leading many notable Houses, including Oxford University Press and Macmillan, he was truly 'an old dog learning new tricks' as we swept through these various language editions.

The Advertising Industry: Real-World Applications

As CEO of an AI-powered advertising company, I've observed how the most interesting developments aren't about automation but about deeper creative understanding in communications. All of these examples contain Collaborative Creativity.

Burger King's *2023* campaign in Brazil celebrated AI's imperfections rather than hiding them. The campaign featured AI-generated Whopper images that looked wrong - cheese dripping impossibly, lettuce defying gravity. The tagline "The Whopper is too iconic to be replicated by AI" turned technical limitations into brand authenticity.

Volkswagen's *2024* ID.4 campaign created ads that wrote themselves based on driving data, but the creativity lay in the human framework that made data meaningful. Different journey types - commutes, weekend drives, long trips - were assigned emotional territories and language palettes. The AI composed unique messages, but humans ensured they resonated emotionally.

Nike's Run Club app uses AI to provide personalised coaching and insights based on running data. The human insight establishes motivational frameworks - the "comeback kid," the "night owl," the "milestone crusher" - while AI handles the scale of delivering personalised experiences to millions of runners.

The Future of the Duet

This creative partnership between humans and AI doesn't just change how we create. It changes who gets to create. By democratising access to high-level research, iteration, and distribution, it empowers new voices to bring their ideas to the world. The grandmother in Lagos with stories to tell, the teenager in Kolkata with game ideas, the engineer with tales from Egypt, the snake-catcher from Florida - all can now access the tools to transform their creative sparks into expressions.

As I've argued, this isn't anthropomorphism, where we treat the machine as human. It's something far more profound: a state of Panthropism. In our dialogue with an AI partner, we're communing with the whole of human endeavour. We have a personal connection to the entirety of our civilisation's knowledge, its art, its science, its failures and its triumphs.

Chapter 4

Creative Industries on the front foot

A Tour of the Future, Happening Now

In the last chapter, we established a framework for our duet with an AI partner: the creative loop of Spark, Draft, and Polish. It's a clean, useful model, but it remains theoretical until we see it 'out in the wild'. Now, we leave the workshop and step out into the world. The hum of this new machine is no longer a distant sound; it's the rhythm section in studios, agencies, and ateliers across the globe.

What follows isn't an encyclopaedia. It's a tour, a trip into the very near future, a future that's already showing its face today. We'll explore different creative domains to see how this new partnership is fundamentally altering three key dimensions of creative life: the workflow, the aesthetic, and the flow of money.

You'll see a unique reflection of our central theme: the intuitive, messy, brilliant spark of human creativity being amplified, challenged, and enabled by the facilitating power of our dynamic, yet inorganic, partner. You'll see how the dance between human and machine isn't leading to a world of sterile, computer-generated uniformity. Instead, it's giving rise to an explosion of new styles, new voices, and new ways of making a living from our innate human desire to create.

Our first stop is in one of the oldest of all human domains: the world of stories.

Section 1: Crafting Our Stories

Publishing, Film, and Games

Narrative is perhaps our oldest technology. From fireside tales to sprawling cinematic universes, we've always used stories to make sense of the world and our place within it. For millennia, the craft of storytelling was bound by the physical limitations of the medium: the time it took to write a book, the immense cost of producing a movie, the painstaking process of coding a game. Now, the AI partner has arrived, and it's acting as a lubricant for the gears of narrative production, allowing us to tell more personal stories than ever before.

Publishing and Literature

For centuries, the author was a lonely figure, wrestling with the blank page in isolation. That era is over, if an author wants it to be. Today, the writer's room can be a team of one human and one AI, a partnership that redefines the writing process from beginning to end.

The Writer's New Room

The **workflow change** is profound. Chen Qiufan, the Chinese science fiction author, used AI systems to generate speculative future scenarios based on current technological trends in Asia for his novel "*AI 2041*" (co-written with Kai-Fu Lee). But crucially, he then applied his human understanding of Chinese culture, philosophy, and social dynamics to shape these scenarios into emotionally resonant stories. The AI provided the technical extrapolation; Chen provided the soul.

In the US, **Author's Guild member K Allado-McDowell** has pushed this collaboration with "*Pharmako-AI*," written in direct dialogue with GPT-3. The text alternates between human and AI passages, creating what they call a "cognitive duet." The result isn't just a book but a new literary form that makes the collaboration itself visible.

This new workflow is giving rise to an **aesthetic shift**.

Serialised fiction platforms are seeing writers use data analytics to understand reader engagement. Author Hugh Howey, who found success with his "*Wool*" series, has publicly discussed using Amazon's data on where readers stop reading to identify pacing issues. He's written extensively on his blog about how analytics helped him understand that his strength lay in apocalyptic tension rather than character backstories, leading him to restructure later works. While not AI-specific, this demonstrates how authors are using data to understand their own creative strengths.

Kindle Vella provides authors with detailed engagement metrics showing exactly where readers stop reading or skip ahead. Several authors have publicly discussed on Twitter and writing forums how they use these insights - not to pander, but to understand which elements of their craft resonate most strongly with readers.

The **flow of money** is adapting globally. **Kindle Vella** in the US has created a micro-payment model where readers unlock episodes with tokens. AI helps authors optimise cliffhangers and pacing, but success still depends on the human storytelling instinct.

Film, Television, and Streaming

The movie industry, with its colossal budgets and armies of specialised craftspeople, is experiencing fundamental shifts through AI collaboration.

Cinema's Collaborative Evolution

The **workflow flip** is dramatic everywhere, but implemented differently by region. In Hollywood, **Waymark** made "*The Frost*," a science-fiction short film. Led by director Josh Rubin, the creative team employed text-to-image AI models, primarily DALL-E 2, to generate the film's visuals. They crafted the prompts fed to the AI and then sifted through thousands of generated images. This intensive curation was essential to piece together a coherent and emotionally resonant narrative from the AI's output. The resulting aesthetic of the film has been described by those involved as "strange" and "otherworldly."

Industrial Light & Magic in the US has developed **StageCraft**, the AI-enhanced virtual production system used in "*The Mandalorian*". It generates photorealistic environments in real-time, allowing directors to shoot "on location" in alien worlds while never leaving the studio. The AI handles the complex calculations of parallax and lighting; the human filmmakers make the creative decisions about mood and framing.

The *2023* sci-fi film ***The Creator*** is an example of modern, AI-facilitated filmmaking. Instead of using expensive green screens, the director shot the movie in real locations across Asia. Afterwards, visual effects artists added all the futuristic robots and ships directly onto that real footage. They used AI-powered tools to help blend these digital effects with the live-action shots, allowing them to create spectacular visuals for a much smaller budget than a typical blockbuster.

The **aesthetic shift** toward personalisation has different cultural expressions. Disney's approach to trailer personalisation reveals how AI enhances rather than replaces human creativity. Human editors craft multiple versions of each trailer - some emphasising action sequences, others highlighting character development or emotional moments.

AI's role comes in the distribution phase: sophisticated algorithms analyse viewing patterns and preferences to match viewers with the trailer version most likely to resonate with them. An action film enthusiast might see explosions and chase scenes, while someone who favours character dramas receives a version showcasing relationships and dialogue. The creative decisions remain entirely human; the AI simply ensures those decisions reach their ideal audience.

The **flow of money** in film and television production is shifting as streaming platforms increase their content investments and turn to AI for competitive advantage, both creatively and in terms of returns on capital deployed. **Netflix** invested approximately $17 billion in content in *2023*. Though their specific allocation to AI-enhanced production tools isn't publicly disclosed, production companies worldwide are openly committed to exploring AI tools for pre-visualisation and other processes to reduce costs.

Games and Interactive Worlds

Video games are experiencing perhaps the most radical transformation globally, with innovations emerging from both AAA studios and indie developers.

The Ultimate Sandbox

The **workflow flip** centres on procedural content generation, but with regional flavours.

Ubisoft's La Forge lab in France developed Ghostwriter, an AI tool that generates first drafts of Non-Player Character (NPC) background dialogue - the ambient chatter you hear from crowds and minor characters. Narrative designers then review and refine these "barks" to fit each game's context. This allows human writers to focus on core story elements while AI handles the volume of incidental dialogue needed for immersive game worlds.

Epic Games has democratised game creation through Unreal Engine's powerful procedural generation tools, which use AI principles to help developers create vast, detailed worlds efficiently. These tools follow rules that create coherent, believable spaces. However, for *"Fortnite"* specifically, the competitive maps are meticulously designed by human level designers - AI is used for character bots and NPCs, not for generating the play areas themselves.

Game developers worldwide use AI and procedural tools to build massive worlds more efficiently. At companies like South Korea's Pearl Abyss (makers of "Black Desert"), these tools help populate environments by placing assets like trees and foliage according to pre-defined rules. The distinctive cultural identity of game regions - whether Korean, Arabic, or otherwise - comes from human art directors and designers who research and interpret cultural elements, not from AI generation.

The **aesthetic shift** toward "living worlds" is exemplified by **Bethesda Game Studios'** Radiant AI system, first introduced in "*The Elder Scrolls IV: Oblivion*" and refined in subsequent games. This system creates NPCs with daily routines - sleeping, eating, and working on schedules - making the world feel more alive.

The **flow of money** has transformed globally. **Roblox Corporation** has built a multi-billion-dollar platform where players become developers, creating games using the platform's scripting and building tools. The platform demonstrates how democratising game creation tools can create new economic opportunities. Some, though not all, of this is AI-enabled.

Section 2: Designing Our Senses

Music, Art, and Fashion

Our senses are the gateway to our emotions. Art, music, and fashion are frequent languages we use to speak directly to them.

The global expansion of AI collaboration in these fields reveals how different cultures are using technology to amplify their unique aesthetic traditions while also creating new universal languages.

Music and Audio

Music was one of the first creative fields to embrace AI collaboration, with innovations emerging from both established music capitals and new creative centres.

The Global Sound Laboratory

The **workflow** transformation is exemplified by **LANDR**, the AI-powered mastering service that analyses millions of professionally mastered tracks to provide automated mastering. Founded in *2014*, LANDR uses machine learning to apply Equalisation (EQ), compression, and other mastering techniques, making professional-quality mastering accessible to independent artists who previously couldn't afford professional mastering engineers. The company has publicly shared details about their AI approach and has mastered millions of tracks.

iZotope's AI-powered assistants in their Ozone and Neutron software analyse audio and suggest starting points for mixing and mastering based on machine learning from thousands of professional productions. The company has been transparent about their AI development, with their "Track Assistant" feature helping producers achieve professional-sounding results more quickly.

Taryn Southern released "*I AM AI*" in *2017*, one of the first albums composed with AI tools including Amper Music, IBM

Watson Beat, and Google's Magenta. She's been remarkably transparent about her process, describing in interviews how she would provide the AI with parameters like tempo, mood, and style, then curate and arrange the generated musical elements. Southern, who had limited formal music training, has explained that AI allowed her to express musical ideas she couldn't execute alone.

The album garnered significant attention not just for its novelty but for raising fundamental questions about authorship and creativity in music. She continues to advocate for AI as a democratising force in creative industries, regularly speaking at conferences about her experience as an early adopter of AI music tools.

The **aesthetic shift** varies by culture. **Brian Eno** pioneered "generative music" - compositions that evolve and never repeat, created through algorithmic systems following rules he defines. His installations and apps like *"Bloom"* (created with Peter Chilvers) generate unique ambient soundscapes that ensure no two listening experiences are identical.

The **flow of money** in music is evolving with AI integration. **Spotify** launched its AI DJ feature in *2023*, using OpenAI technology to create personalised music selections with AI-generated commentary voiced by Spotify's Head of Cultural Partnerships, Xavier "X" Jernigan. The feature aims to increase user engagement and time on the platform.

Music education is seeing AI adoption through platforms like Yousician and Simply Piano. They use AI to provide real-time feedback to students learning instruments, and demonstrate how AI is creating new revenue streams in music education.

Fine Art and Illustration

The art world's engagement with AI has produced some of the most culturally distinctive applications globally.

Refik Anadol creates what he calls "data sculptures" and "data paintings." His installation "*Unsupervised*" at MoMA (*2022-2023*) used machine learning to process and visualise the museum's collection of over *200,000* works, creating constantly evolving digital animations that he describes as the museum "dreaming". The work explores connections between artworks across MoMA's collection, making visible the invisible relationships between different periods and styles.

The **aesthetic shift** is toward participatory digital art. **TeamLab** in Tokyo creates installations where AI responds to visitors' movements and touch, generating unique patterns and colours that follow each person through the space, making every visitor a co-creator of the experience.

The **flow of money** is transforming. **Christie's** auction house sold Beeple's "*Everydays: The First 5000 Days*" for $69 million, legitimising digital art. But more interesting is **Art Blocks** in the US, where collectors purchase generative algorithms that create unique artworks at the moment of minting. In this instance, we see artists selling systems, not objects – this is a whole new world of creativity.

Fashion and Digital Wearables

Iris van Herpen uses AI and 3D printing to create her avant-garde collections. She collaborates with architects and scientists, using algorithms to generate complex structures that would be impossible to create by hand, translating digital designs into physical garments that look like wearable sculptures.

The Fabricant in the Netherlands exists entirely in digital space, creating high-fashion garments that will never be physically produced. Their "Iridescence" dress sold for $9,500 despite existing only as data, pioneering fashion freed from physical constraints.

The **aesthetic shift** moves toward "responsive fashion". **Google's** Project Jacquard weaves conductive fibres into garments, creating clothes that sense touch and movement. Their collaboration with Levi's produced a jacket that lets cyclists control their phones through sleeve gestures and receive navigation alerts while riding. It's fashion as protective interface.

The **flow of money** is bifurcating. Physical fashion sees designers using AI as a creative partner, such as **Christopher Raeburn** who uses AI to generate zero-waste pattern layouts from his sketches, with the algorithm suggesting unexpected ways to use every scrap of fabric. On the other hand, digital fashion creates entirely new markets. **The Fabricant,** which I previously mentioned, uses AI to create impossible materials like "liquid metal" fabrics that respond to virtual physics (by which I mean, it models how Physics in the real world would affect the garments), selling these purely digital garments for thousands of dollars.

Section *3*: Shaping Our Attention

Advertising and Photography

In a world saturated with information, the most valuable commodity is attention. The creative industries dedicated to capturing it are undergoing fundamental rewiring globally.

Advertising and Branding

The advertising revolution through AI manifests differently across markets but shares common themes.

Ogilvy Paris created an AI system for client IBM that writes ads showcasing AI capabilities. The twist: the AI explains its own creative process in the ads, demystifying AI for business audiences. Human strategists ensure the tone is approachable, not intimidating. It's AI making itself understandable through human guidance.

Wieden+Kennedy worked with KFC to create AI-generated ads featuring Colonel Sanders. Human creatives defined the brand voice and humour, then collaborated with AI to generate scripts. The result felt authentically KFC while exploring new creative territories suggested by AI.

BBDO used AI to create personalised Snickers ads. Human creatives established the "You're not you when you're hungry" concept, then AI analysed social media to identify real-time "hungry moments" and generate relevant responses. Humans curated the outputs to ensure brand safety and humour.

Goodby Silverstein & Partners has experimented with AI tools for ideation and concept development. Creatives use AI to generate initial concepts and variations, but human judgment determines which ideas capture the brand's spirit and cultural relevance. The agency views AI as an ideation accelerator, not a replacement for human creativity.

The **aesthetic shift** can be seen in "liquid branding". **Coca-Cola's** "Create Real Magic" campaign invited global audiences to remix brand assets using AI. Times Square displayed the best creations, making consumers co-creators of the brand's visual identity. This was a playful, mass interaction that gave Coke creative leadership and its consumers a sense of individualised engagement in a new technology.

The **flow of money** has transformed. Traditional agency retainers are being replaced by performance-based creative models. **WPP** partnered with AI companies like Synthesia to offer clients AI-powered video creation services, where human strategists design campaigns and AI generates multiple personalised versions. Performance-based platforms now let agencies tie payment to AI-optimised results rather than hours worked, shifting from commissioning campaigns to subscribing to continuous creative optimisation.

Photography and Imaging

Photography's relationship with AI varies dramatically by cultural context and artistic intent.

Mario Klingemann, a German artist, uses neural networks to create portraits that question photographic truth. His *"Memories of Passersby I"* uses AI to generate endless unique faces that never existed, with Klingemann curating the training data and refining the algorithm to achieve his aesthetic vision of uncanny near-humanity.

Es Devlin collaborated with AI to create her *"PoemPortraits"* project. Visitors contribute a word, AI generates unique verses, and these are combined with photographic portraits. She provides the concept and curation; AI enables mass personalisation of poetic imagery.

The **aesthetic shift** is toward "computational photography". **Google's** Pixel phones use AI to capture images impossible with traditional optics - seeing in near darkness, removing unwanted objects, creating depth from single lenses. Every smartphone photographer now wields computational superpowers.

The **flow of money** is disrupting traditional models. **Getty Images** has partnered with NVIDIA to create licensed generative AI tools, with stated commitments to share revenues with contributors, though the specific compensation mechanisms remain to be fully detailed.

Meanwhile, platforms like **DALL-E** and **Midjourney** create new markets for "prompt photographers" who specialise in crafting words that generate specific visual outcomes. This isn't just typing "make me a pretty picture" - it's become a sophisticated craft. The best prompt artists understand how these models interpret language: they know that adding "volumetric lighting" creates depth, that referencing specific artists or photographic techniques triggers certain aesthetics, that technical photography terms like "85mm lens" or "golden hour" produce predictable effects.

Marketplaces like PromptBase have emerged where prompt engineers sell their carefully crafted instructions. Some specialists work as consultants, helping brands develop AI-compatible visual languages. Online communities share prompt libraries and techniques, building a collective understanding of how to "speak" to these models.

The skilled practitioners layer prompts methodically: starting with subject and composition, adding lighting cues, incorporating artistic references, fine-tuning with technical parameters. They understand that subtle word choices dramatically alter outputs. It's photography without cameras - but it still requires an eye for composition, an understanding of light, and the ability to envision what doesn't yet exist.

In every domain, the story is consistent but culturally varied. The duet between human and machine creates not homogeneous global culture but amplified local voices. London's wit, Lagos's energy, Tokyo's precision, São Paulo's warmth - all find new expression through AI collaboration. The future of creativity isn't singular but munificently plural, with each culture using these tools to tell their own stories more powerfully than ever before.

Chapter 5

Beyond the Arts

Creativity is Everyone's Business

So far, we've looked primarily at the arts. We've explored how the human-AI duet is reshaping the work of the novelist, the musician, the filmmaker, and the fashion designer. But to confine our understanding of creativity to these domains alone would be to miss the bigger picture. It would be like admiring the beauty of a single, brilliant wave and ignoring the tide that lifts it.

Creativity isn't the exclusive property of those we call 'artists'. It's the foundational human response to a challenge. It's the spark that drives the scientist to question a given truth, the engineer to find a more elegant solution, the urban planner to imagine a more liveable city, and the chef to discover a new combination of flavours. As I've argued, every one of us possesses this originative force. The creative pulse of Earth has a mighty number of beats.

The expression of that pulse in many fields has been constrained by the immense 'craft' of managing complexity. The scientist wrestles with colossal datasets, the architect with intractable physics, the doctor with a universe of medical literature. These are domains where the analytical work is so overwhelming that it can stifle the very intuition it's meant to serve.

This is where our new partner, the Allied Intelligence, changes the game. By taking on the heavy lifting of computation, analysis, and simulation, AI is setting the stage for a dramatic emancipation of creativity in fields far beyond the arts. It's allowing human ingenuity to be applied directly to our most urgent and complex problems.

In this chapter, we'll visit the laboratory, the design firm, and the city square. We'll see how the same creative duet we explored in the arts is now playing out across the grand challenges of our time, organised into three fundamental human endeavours: how we design the world around us, how we solve the problems of our survival, and how we enhance the fabric of our daily lives.

Designing Our World: Architecture, Products, and Cities

The world we inhabit isn't an accident. It's a product of design. From the chairs we sit on to the cities we live in, every human-made object and space is the result of a creative act, admittedly some more deliberate than others. It's an attempt to solve a problem: how to provide shelter, how to make a tool more useful, how to make a community more connected. This is a domain where the creative vision must constantly negotiate with the unyielding laws of physics and the complex needs of people.

Architecture

The collaboration in architecture is now about creating spaces that are more attuned to both the environment and their human inhabitants.

Zaha Hadid Architects (ZHA) in London uses AI-driven generative design extensively. For the Beijing Daxing Airport, their CODE group used parametric design tools to generate and test thousands of configurations. AI optimised passenger flow patterns for *100* million annual travellers while human architects shaped these solutions into flowing spaces that feel like landscapes, reflecting Chinese concepts of continuous movement.

The workflow flip is moving from static blueprints to dynamic, generative design. **Autodesk's** generative design tools, used by architects worldwide, allow designers to input goals and constraints - like maximising daylight while minimising heat gain. The AI generates thousands of options. Human architects then select and refine designs based on aesthetic and cultural criteria.

Architechnologies in Ontario uses AI to generate floor plans based on client requirements. Architects input parameters like number of rooms, square footage, and site constraints. The AI produces multiple layouts optimising for factors like natural light and traffic flow. Human designers then refine these, adding the cultural and emotional elements that make spaces liveable.

The aesthetic shift shows in projects using environmental data to shape form. **NBBJ** architects used AI analysis of solar patterns to design the Googleplex roof canopies. The AI calculated optimal angles for shade throughout the day; human designers turned these into sculptural forms that create dynamic light patterns.

Product and UX Design

Adobe's Sensei AI assists designers throughout the creative process. In Adobe Illustrator, designers can sketch rough shapes and Sensei suggests refined versions. The human maintains creative control, selecting which suggestions capture their vision and modifying them further. This collaboration speeds up technical execution while preserving artistic intent.

Autodesk's Dreamcatcher (now part of Fusion *360*) enables generative design for products. Designers input requirements - like "a chair that supports *300* pounds using minimal material." The AI generates hundreds of organic-looking solutions. Human designers select based on aesthetics and manufacturability, often discovering forms they wouldn't have conceived alone.

The Grid was an early AI website builder where users provided content, and AI designed layouts. While the company ultimately failed, it pioneered the concept of AI as a design partner - users described their vision in natural language, the AI generated designs, and humans refined the results.

Airbnb uses AI to help hosts optimise their listings. The AI analyses millions of successful listings to suggest photo angles, description improvements, and amenity highlights. Human hosts decide which suggestions align with their property's unique character.

Netflix's recommendation system is a human-AI collaboration. Human editors create thousands of micro-genre tags ("Romantic Foreign Movies featuring a Strong Female Lead"). The AI then uses viewing patterns to match users with content. Humans provide the cultural understanding; AI handles the scale.

Drug Discovery and Medicine

Atomwise uses AI for drug discovery through virtual screening. Researchers provide target proteins and desired drug characteristics.

The AI screens millions of compounds, predicting which might bind effectively. Human scientists then select the most promising candidates for laboratory testing. This collaboration has identified potential treatments for Ebola and multiple sclerosis.

Insilico Medicine uses generative AI to design novel drug molecules. Scientists specify the biological target and desired properties. The AI generates new molecular structures that don't exist in any database. Human researchers evaluate these for safety and synthesizability. In *2019*, they designed, synthesised, and validated a new DDR*1* kinase inhibitor in just *46* days.

IBM Watson for Oncology was trained on Memorial Sloan Kettering's treatment protocols. Oncologists input patient data; Watson suggests treatment options based on similar cases. Doctors make final decisions, using Watson's analysis to ensure they haven't missed options.

It faced several significant criticisms. According to *2018* reporting by STAT News based on internal documents, Watson sometimes gave questionable treatment recommendations. Major hospitals including MD Anderson Cancer Center and Denmark's Rigshospitalet discontinued their Watson programs after spending millions. A key limitation was that Watson was trained primarily on Memorial Sloan Kettering protocols, making it less applicable to hospitals with different treatment approaches or patient populations. Studies found Watson largely confirmed what oncologists already knew rather than providing novel insights. Despite these challenges, Watson for Oncology did help establish the template for human-AI collaboration in healthcare.

BenchSci uses AI to help researchers find antibodies for experiments. Scientists describe their experimental needs; the AI searches millions of papers to find antibodies used in similar contexts. Researchers then select based on their specific requirements. This has reduced average antibody selection time from weeks to minutes.

Climate and Sustainability

Climate TRACE, co-founded by Gavin McCormick, uses AI to track global greenhouse gas emissions. The AI analyses satellite imagery to identify emission sources. Human experts validate findings and work with governments and companies on reduction strategies. This collaboration provides emissions data for countries that lack monitoring infrastructure.

Microsoft's AI for Earth provides cloud computing and AI tools to environmental researchers. One project with SilviaTerra uses AI to analyse satellite imagery to assess forest carbon storage. Foresters provide ground truth data; the AI extrapolates across millions of acres. Human experts interpret results for conservation planning.

The Ocean Cleanup uses AI to identify plastic pollution in rivers. Cameras capture images; AI identifies and counts plastic items. Human researchers use this data to position cleanup systems effectively. This collaboration helps prioritise cleanup efforts in the most polluted waterways.

Social Innovation

Crisis Text Line uses AI to help counsellors prioritise messages. The AI analyses incoming texts for crisis indicators. Human counsellors receive alerts about highest-risk conversations. This collaboration helps ensure those in immediate danger get help first while maintaining the human connection essential for crisis support.

Khan Academy uses GPT-4 to power Khanmigo, an AI tutor. Students work through problems with AI guidance. The AI doesn't give answers but asks Socratic questions. Human teachers monitor progress and intervene when needed. This collaboration provides personalised tutoring at scale while maintaining human oversight.

Creative Industries

RunwayML provides AI tools for video creators. Editors can use AI to remove backgrounds, generate animations, or create effects. The human provides creative direction; the AI handles technical execution. Films like *"Everything Everywhere All at Once"* used Runway for some visual effects.

Boomy allows users to create music with AI. Users select style and mood; the AI generates instrumental tracks. Humans then edit arrangements, add vocals, and refine the mix. Over *14* million songs have been created this way, with some artists earning streaming revenue.

Canva's Magic Write uses AI to help users create marketing copy. Designers input their goals and key points; the AI generates multiple text options. Users select and modify the copy to match their brand voice. This collaboration helps non-writers create professional marketing materials.

Manufacturing and Industry

Siemens uses AI in their factories for predictive maintenance. The AI analyses sensor data to predict equipment failures. Human technicians receive alerts and perform maintenance before breakdowns occur. This collaboration has reduced unplanned downtime by up to 50% in some facilities.

General Motors works with Autodesk's generative design for car parts. Engineers specify requirements like strength and weight limits. The AI generates organic-looking designs that use less material. Human engineers select designs that can be manufactured efficiently. This produced a seat bracket that's 40% lighter and 20% stronger than the original.

Architecture firm **Spacemaker** (acquired by Autodesk) uses AI for site planning. Architects input site constraints and building requirements. The AI generates layouts optimizing for sunlight, noise, and views. Human planners select and refine options based on community needs and aesthetic goals.

These examples demonstrate genuine human-AI creative collaboration where humans provide vision, values, and judgment while AI provides analytical power, pattern recognition, and the ability to explore vast solution spaces. The key is that neither human nor AI creates alone - the outcome emerges from their creative collaboration.

Part III: Momentum

The unstoppable force meets the moveable mind

Chapter 6

The New Creative Alliance

Dispatches from the Age of Mass AI

The moment the first public generative models opened their gates in November *2022*, it felt less like the release of a new app and more like the breach of a dam. Water that had been accumulating for seven decades of computer science suddenly surged into every valley of human imagination.

In the cafés of Lagos, in the loft studios of Reykjavik, in the dorm rooms of São Paulo, artists and engineers, marketers and poets discovered that sentences could now paint pictures, that harmony could be coaxed from statistical noise, that design space itself had grown new dimensions overnight. The sensation wasn't one of incremental progress. It was the shiver we associate with plate tectonics, with Gutenberg's first sheet slipping out of his press, with the copper threads of the first telephone carrying a human voice across a continent.

For the past two chapters, we've travelled quickly across that altered topography, noting the new ridgelines in publishing, film, science, and industrial design. Yet such sketch-maps can flatten lived experience. To grasp the emotional and intellectual scale of the change we need to slow down, to stand inside the places where humans and machines now share the bench.

The following pages linger there, listening for the click of intention meeting algorithm. They're arranged not as a catalogue but as a braided meditation on two emergent modes of creativity.

The first mode might be called **Collaborative Creativity**. A human, restless with vision, encounters a technical barrier and invites an artificial partner to overcome it. The person supplies context, taste, risk, the sense of what it means to matter. The machine supplies a patience beyond fatigue, a willingness to wander a billion blind alleys in microseconds, and a capacity for pattern recognition that borders on clairvoyance. Together they form something neither could achieve alone. We've watched such duets before - the choreographer with her trusted rehearsal pianist, the playwright with his dramaturg - but never at this scale, never with a partner whose memory contains so large a slice of human culture.

The second mode is less familiar and more disquieting. I'll call it **Emergent Creativity**. Here the machine is handed an objective but no recipe. Instead of following the grooves of human precedent it wanders wilderness regions of mathematics or chemistry or language, and returns with artefacts we didn't envisage or anticipate. They're novel, useful and sometimes inexplicable, as though pulled from an unfamiliar pocket of space. Whether we choose to name this behaviour creativity, invention or simply accident remains a live debate, but the debate itself is evidence of a shift in the idea of agency.

Let's move, then, from abstraction. What follows is an interwoven narrative, flowing back and forth between the collaborative and the emergent, because that's how the real world now operates. On some mornings, a songwriter asks the AI to separate a grainy voice from a piano; on others, it wakes the scientist with an email announcing a compound that might cure a hospital superbug. The boundary is porous. The stories accumulate. And each story is, in its way, a small rehearsal for the larger conversation we must hold about authorship, ownership and the future shape of genius.

Collaborative Creativity: The Human Hand on the Tiller

The Beatles' Final Song

The instinct to finish unfinished business may be the most human of impulses and nowhere is that yearning clearer than in the often referenced saga of *Now and Then*. Paul McCartney's wish to complete John Lennon's late-night piano demo began as a private ache, the kind of half-spoken longing that haunts many survivors of partnership.

The cassette Yoko Ono handed over in the early *1990*s contained not only a skeletal song but the ambient hiss of a Midtown apartment and the brittle clank of the piano itself. George Martin's engineers, armed with what were then state-of-the-art filters, tried to pry the frequencies apart like antique conservators teasing pigment from varnish. Each attempt left Lennon's voice sounding like a ghost rattling a tin cage. With Harrison's taciturn shrug the project was shelved. The fragment joined the melancholy shelf of artefacts labelled *impossible*.

Three decades later, Peter Jackson's obsession with his documentary *Get Back* led his team to train a bespoke neural network that could hear not just the difference between mastoid and mandible resonance but the minute timbral quirks of each Beatle's guitar pick. The neural network became a forensic ear able to lift the creak of Ringo's drum stool from the murmur of a roadie's cough. McCartney, watching test reels, recognised it. The mix-mastering AI was pointed at Lennon's cassette. Hours later a clean vocal, vulnerable and immediate, rose from the speakers. It wasn't quite time travel, but it was close enough that McCartney felt the peculiar discomfort of the bereaved meeting the departed in a well-lit room.

From that point, the collaboration reverted to analogue intimacy. McCartney laid down a Hofner bass line built to converse with Lennon's right-hand arpeggios. Ringo tracked a drum performance with brushes that refused to overpower the voice. Archived Harrison guitar textures, themselves a relic of a failed nineteen-ninety-five session, were woven in, so that all four Beatles were literally (and not so literally) playing together. Giles Martin conducted a small string section, quoting voicings his father had once scribbled in the margins of *Eleanor Rigby*.

The AI's job was finished; it sat politely in the corner, like a master restorer who's lifted centuries of grit from a fresco and now watches pilgrims weep before it. The resulting single is less a novelty than a compressed parable: the human heart carries the flame; the machine tends the lamp glass. (Personally, I find it a shame that all this dedication resulted in a track we'd probably have thought barely worthy of the Beatles had they all been alive and releasing it, but... there's romance in the story and in the finishing of business.)

Film Directors Meet Sora

In cinematography, five directors were invited by Tribeca Film Festival in *2024* to create short films using OpenAI's text-to-video engine, Sora. The selected filmmakers - Bonnie Discepolo, Ellie Foumbi, Nikyatu Jusu, Reza Sixo Safai, and Michaela Ternasky-Holland - each brought their unique creative vision to explore this new technology.

The shorts premiered at Tribeca *2024*, demonstrating how filmmakers could use AI-generated video as a creative tool. Directors worked with Sora by crafting text prompts to generate footage, then curating and editing the results to realize their artistic visions. This collaboration represented one of the first public showcases of professional filmmakers using AI text-to-video generation for narrative storytelling.

The Virtual Pop Star

Music, memory and image coalesced in an innovative collaboration when Ash Koosha created his virtual pop star YONA using AI. The British-Iranian electronic musician used AI to generate YONA's voice, appearance, and even personality. Working with speech synthesis and 3D modelling tools, Koosha created songs where YONA performs vocals that were generated by AI trained on human singers.

For live performances, Koosha appears on stage while YONA is projected as a hologram, with her movements and vocals controlled by AI responding to the music in real time. The project explores questions of authenticity and authorship in music - audiences connect emotionally with YONA despite knowing she's entirely artificial. Koosha describes it not as replacing human creativity but as exploring new forms of musical expression that wouldn't be possible without AI collaboration. The project has been featured in major outlets including Wired and has performed at venues like the Barbican Centre in London.

The Rapid Evolution

Stories like these proliferate. And it develops by the day. I read so many newsletters, articles, papers, and even books on AI that it's come to the stage where I often 'put them all into an AI' and ask it to inform me of the most material developments or advances that have taken place overnight.

Even though I think as widely as I can about definitions of creativity, sometimes I just don't know what to make of the next major advance I see. Rowan Cheung, who runs a particularly useful daily newsletter on AI on LinkedIn, The Rundown, posted recently:

"BREAKING: OpenAI just launched ChatGPT Agent. I had early access, and the agent built me a complete early retirement plan: > Found local tax laws (Vancouver) > Analyzed average monthly spend rates > Calculated savings needed to retire at 30 > Researched optimal investment allocations > Found tax optimization strategies I'd never heard of > Created a downloadable presentation with results. This

would've cost me $5,000+ from a financial advisor and taken weeks. The new feature essentially allows ChatGPT to think, plan, and execute complex tasks on its own virtual computer while you do other things. It does this by bringing these three ChatGPT capabilities into one unified system: - Operator's web interaction skills - Deep Research's information synthesis - ChatGPT's conversational intelligence. I think with ChatGPT Agent now, and especially as it gains access to more tools, we're finally going to see the rise of a new AI skill category in Agent Management. Agents are finally becoming capable of doing real work autonomously, so anyone who learns how to effectively orchestrate agents will have a huge advantage."

Can one quiver with excited anticipation that we might even be seeing the harbingers of a new world, where humans no longer need to 'work' - that is, the daily grind, to turn up repetitively and tread the mill - and instead we may be looking at the dawn of human volition being executed by our Allied Intelligence? I'm not an advocate of the 'we should not work if we don't have to' mentality, but I wonder whether we are seeing the glimmer of the dawn of a substantially different world economy, in which much of human volition is outsourced to AI for execution.

The New Aesthetic: What We're Getting Used To

The Evolution of Visual Grammar

Several months ago, I overheard some seasoned book publishers observe, "It's nice, but it looks... too AI." They were looking at some new advertising for a title.

Translation: the piece wasn't fooling anyone into believing it was a photograph. Somewhere in the uncanny swirl of colours or near-subliminal geometry, the eye detected the tell-tale signature of the machine. That signature was seen as a fault.

It made me think.

Civilisation has lived through this before. Every era births a new visual grammar that initially offends prevailing taste, only to become tomorrow's mainstream. Sometimes this results from new techniques, sometimes from new technologies. The status quo in our aesthetic ecosystem is regularly disrupted, and before we habituate to its new parameters, we normally rail against them for a while. Re-acquainting ourselves with earlier artistic scandals can help us to see the so-called "AI aesthetic" not so much as a glitch but as the next chapter in the evolution of style.

Historical Precedents

Remember when "unfinished" was an insult? In Paris in 1874 Monet, Degas, and friends mounted the first Impressionist exhibition after being rejected by the Salon jury. Critics sneered that their canvases were mere "sketches", branded the movement with the dismissive slur "Impressionism", and complained about "unfinished or even sloppy" brushwork.

Van Gogh was generally seen as a madman, in his time. He sold exactly one painting in his lifetime, *The Red Vineyard (1888)*. Just linger on that (initial) meagre recognition. His abrasive textures, exaggerated colours, and swirling skies scandalised his neighbours. Later, they defined post-Impressionism and, indeed, the gift-shop poster industry. My children's bedrooms were festooned with prints of such works, as part of my effort to familiarise them with artistic talent.

Long before generative models, photography itself was denounced as art's "mortal enemy." Poet Charles Baudelaire wanted the new medium banished to science textbooks. In *1913*, critic Marius de Zayas thundered: "Photography is not Art. It is not even an art... (just) the plastic verification of a fact." But of course, without the technology of cameras and films, we wouldn't have treasured Man Ray and his surrealist and avant-garde photography - nor Ansel Adams with his famously 'Straight Photography and Zone System' that stylised 'un-natural', high-contrast, sharply detailed black-and-white landscapes.

The Pattern Repeats

The auditory arts reveal the same pattern. When Stravinsky's "*The Rite of Spring*" premiered in *1913*, it caused a riot in the concert hall. It was denounced as "barbaric" and "savage noise". Today, it's standard repertoire in symphony orchestras. What was first seen as cacophony is now a classic.

The electric guitar was initially dismissed as a gimmick. It wasn't acceptable as a "real" instrument for serious musicians. Early rock and roll was condemned as "the devil's music". Electronic music pioneers like Kraftwerk were mocked for replacing "authentic" instruments with synthesizers. Each innovation first shocked, then normalised, then influenced everything that followed.

Remember the mockery of Auto-Tune? It evolved from a corrective production tool to a deliberate aesthetic choice that defined *2000s* hip-hop and pop. What was once derided as "cheating" was then deliberately flaunted by Cher's *"Believe"* (*1998*).

Embracing the AI Aesthetic

So, what counts as 'too AI'? Right now, generative images wear their origins on their sleeve: hyper-sharp fabrics, impossibly even lighting, cosmic depths-of-field, and other unfamiliar aesthetic markers. We instinctively try to sand these artifacts away to achieve photographic plausibility. In doing so, we risk repeating the mistakes of the past - judging a new medium by the criteria of an old one.

What if the bloom, the gloss, the dream-like morphologies are precisely the core DNA of our new aesthetic, and give, for example, a book advertisement those crucial three seconds of stop-time on a scrolling phone screen.

What if they express emotions in new ways, produce psychological hooks that arrest attention and instigate engagement, in a way not previously experienced? What if, for example, they bring an author's work to life and gain attention in a way previously unimagined? (Spoiler alert - I know for a fact that they do exactly these things. Data from 'AI-aestheticized' advertising at Shimmr AI shows it attracts at least 6x more reader attention than previous benchmarks, and more than 12x conversion to e-commerce interaction.)

The charge that AI is "soulless" echoes previous artistic backlashes. History shows the antidote is authorship. The difference between prompt-spam and purposefully conceived ad campaigns remains a discerning and subtle brief, full of nuance and finesse. AI needs stellar, experienced creative briefing, as much as advertising always has.

Treat AI as a new genre, not cheap labour. Human creativity was the key in unlocking fresh aesthetics in painting, photography and audio-art. 'Crude' brushstrokes gave us new interpretations of our world; photography didn't wipe out painting and 'video didn't call the radio star' to quote the well-known song written by Trevor Horn, Geoff Downes, and Bruce Woolley; it widened the menu; we still sing.

Generative AI imagery sits alongside illustration, collage, and photography, giving marketing departments one more dialect of visual language. It goes without saying that it should be co-created by human artists. Seen this way it's a transformation, an evolution, and an exciting one.

The pattern is clear: every new tool first looks like a toy, then like a threat, and finally like a positive choice of style. Readers have already seen a lifetime of photo-realism. They might be ready, even eager, for new aesthetics.

Let's not wait fifty years to recognise the beauty in the machine's brush, moved by the human mind.

Emergent Creativity: Sparks in the Dark

Collaborative stories feel safe because each participant's role is clear. The newer, machine-led examples of Emergent Creativity feel uneasy because those roles mix together.

We usually say creativity requires conscious intent, a tangled personal back-story and the nervous wait for inspiration. Yet machines are beginning to show something that looks a lot like creativity, even though they've no feelings at all. At least, that's what we believe.

Mathematics Reimagined

Matrix multiplication sits at the heart of almost every simulation, graphics program, and neural network. For half a century mathematicians chipped away at the standard method, squeezing out only small speed-ups.

DeepMind tried something different. They cast the search for a faster algorithm as a board game and let an AI agent called AlphaTensor play. The legal "moves" were just the valid steps of linear algebra, and the goal was to finish the calculation in as few steps as possible. Without studying human proofs, the agent tested long chains of operations, millions that'd never been printed in a textbook. After enormous variations it uncovered new, shorter recipes that beat every known approach.

When AlphaTensor's results appeared in *Nature* in October *2022*, mathematicians felt more delight than resentment. Many compared the moment to finding a fresh, easier climbing route up a mountain that they'd spent decades scaling.

Code That Writes Itself Better

A year later DeepMind pointed the same idea at the computer's most basic language: 'assembly'. Assembly is the set of single-step commands understood by a CPU, the chip that does every

The New Creative Alliance

calculation in your laptop or phone. Each command costs time, so a program that needs fewer commands runs faster.

The team asked their new agent, AlphaDev, to rewrite a small but common sub-routine: the sort that orders a handful of numbers. The challenge was to finish the task in as few CPU instructions as physics would allow.

Their yardstick was Low Level Virtual Machine (LLVM), the open-source toolchain that compilers use to turn languages like C++ or Rust into assembly. LLVM's built-in sort routine had been polished by experts for years. AlphaDev, working from scratch and without seeing human examples, generated instruction sequences that the stopwatch showed were faster.

Engineers reviewed the code. It looked unusual but compiled cleanly and produced correct results, so the LLVM maintainers accepted the change. The community now runs millions of machines on assembly that no one fully understands, but everyone trusts because it wins the race. This is, in parts, scary and exciting. Usually, we habituate to things we can understand. In this instance, something not well understood, and created by machines, has become part and parcel of daily life.

Is 'unfathomability' a definition of creativity? The machine had done something a human had never managed to do.

New Materials from Digital Dreams

Chemistry is changing in the same way. DeepMind's GNoME model began by reading every public crystal-structure database it could find, along with several private ones. After learning the patterns, it invented about *2.2* million new inorganic crystal formulas that had never been reported. Computer checks suggested nearly *400,000* should be stable in real life.

A group of materials scientists at Lawrence Berkeley National Laboratory picked seventeen promising candidates and successfully synthesised them in the lab. The atomic lattices matched GNoME's predictions to within a few billionths of a metre. Among the confirmed examples were lithium-rich compounds whose internal channels let ions move quickly, the key to lighter, faster-charging batteries. What normally takes years of furnace work arrived as a batch of ready-made blueprints in a single research sprint.

The Protein Revolution

Proteins, DNA and drug molecules have seen a similar leap. AlphaFold *2* told scientists how a lone protein folds. The newer AlphaFold *3* shows how that protein might pair with RNA, DNA or a small-molecule drug.

Drug-development teams now run their early ideas through the model to see which ones are worth the cost of wet-lab testing. It flags designs that fit the target site well and warns when the same molecule might bind elsewhere and cause side-effects.

One medicinal chemist said using AlphaFold *3* feels like talking to a colleague who's memorised every entry in the Protein Data Bank and can guess the rest on the spot.

The First AI Antibiotic

Perhaps the most headline-friendly example arrived from the labs of MIT. Faced with the dispiriting antibiotic resistance crisis, the researchers trained a model on a modest data set of seven-and-a-half thousand molecules. They asked it to rank which unknown compounds might inhibit a pathogen responsible for stubborn ICU infections.

Within two days, the algorithm nominated a molecule later dubbed halicin (after HAL 9000). In vitro tests showed potent activity against *Acinetobacter baumannii* with negligible harm to human cells. The structural novelty of the molecule startled medicinal chemists; it didn't fit the scaffolds of known antibiotic classes.

The world's first AI-discovered antibiotic isn't yet on pharmacy shelves, but the proof-of-principle reversed the gloomy narrative that the antibiotic pipeline was dry. A machine had found a well in the desert. Is invention 'creativity'?

The Expanding Frontier

More successes appear every month.

In aerodynamics, researchers at NASA Ames used AI to design a wing that reduces drag by 8% while maintaining lift - a combination human designers had tried and failed to achieve for decades. The AI tried millions of simulated wing shapes, eventually producing a design inspired by swift wings with micro-vortex generators that no human had considered. Wind-tunnel tests confirmed the improvement.

In the legal world, the well-regarded law firm Allen & Overy adopted an AI platform named Harvey. Trained specifically on legal data, Harvey acts as an expert assistant, able to analyse complex documents and produce high-quality first drafts of contracts and memos. This allows the firm's senior lawyers to augment their own expertise, significantly increasing the speed and efficiency with which they can handle intricate legal work.

Contours of Authorship and Value

The Philosophy of Machine Creation

Some critics say creativity belongs only to humans because we feel doubt, work under pressure, and know we'll die. In their view, a machine just searches through options without caring about the result.

Yet when a violinist plays Bach we don't disqualify the music because the performer didn't compose it. When AlphaDev writes assembly we must decide whether the absence of pulse and respiration negates the ingenuity of the algorithm.

Law and philosophy will wrestle with that verdict for decades, but in boardrooms and laboratories, the working definition is already pragmatic: if the artefact confers competitive advantage, if it reduces cost or expands possibility, it's welcomed. The question of soul is postponed.

Legal Frameworks Bending

Copyright law is under pressure.

When Coca-Cola projected community-generated posters thirty metres high in Manhattan, who owned the derivative snow-globe Santa drinking in slow motion? The original clip art rested in the Coca-Cola archive. The user typed the prompt. The model interpolated. The billboard vendor profited.

Already we see licencing frameworks that treat the model as a subcontractor: rights flow back to the brand, royalties trickle to the crowd, indemnities crawl up the chain towards the model provider. We're building new plumbing while the water is running.

Academic Attribution

In academia, attribution norms face similar strain. Journal articles now carry paragraphs acknowledging that figures were generated by diffusion models, that statistical analyses were cross-checked by large language models.

The journal *Nature* requires disclosure of AI use in methods sections, while *Science* debates whether AI should be listed as co-author. Some publishers require disclosure, others don't, leading to a patchwork of transparency that recalls the early internet's struggle with hyperlink citation. One can sense the collective scramble for a new etiquette.

I think we'll see a rapid convergence toward standardised disclosure, but with an interesting twist. Within 2-3 years, major journals will likely adopt something like a "Creative Commons" for AI attribution - a set of standardised badges or declarations that specify exactly how AI was used (ideation, writing, analysis, image generation, etc.).

The co-authorship debate will resolve not by listing AI as an author, but by creating a new category entirely - perhaps "AI

Systems Used" alongside "Author Contributions." This sidesteps the philosophical question while maintaining transparency.

What's more interesting is what happens next: I suspect we'll see a brief period where "AI-free" research becomes a badge of honour (like "organic" food labels), followed by acceptance that AI use is so ubiquitous it's only worth noting when it's *not* used.

The real evolution will be in how we describe the collaboration. Instead of just "used ChatGPT", we'll develop nuanced vocabulary: "conceptual dialogue with", "statistical verification via", "prose refinement through", etc. The precision will matter legally and ethically.

Eventually, I think the question will flip entirely. Rather than asking "Did you use AI?" we'll ask, "How did you verify this *wasn't* hallucinated?" or "What was uniquely human about your contribution?" The burden of proof will shift from defending AI use to demonstrating human insight.

Just as we no longer cite spell-checkers or calculators, routine AI use will become invisible. But unlike those tools, AI's capacity for generation means we'll always need some framework for understanding who (or what) contributed which ideas. The etiquette we're scrambling for now will solidify into something as natural as citing sources - necessary, standardised, and ultimately unremarkable.

The Economic Shift

Economically, the most immediate shift is in what I call the **Ratio of Dream to Effort**, or, to coin a term, **RODE**.

Throughout history, this ratio has been crushingly low. A filmmaker might dream of a sweeping epic but face years of fundraising, crew assembly, and production logistics. A musician could hear symphonies in their head but needed an orchestra to realise them. The effort required to manifest a dream was so enormous that most visions died unborn. High RODE meant that only the wealthy, the connected, or the phenomenally persistent could create.

AI fundamentally disrupts this equation. When a single person with a laptop can generate concept art, compose soundtracks, write scripts, and create animated sequences, the RODE skyrockets. What once took *100* units of effort per unit of dream might now take *10*, or even *1*. The dream remains as vivid as ever, but the path to expressing it has shortened dramatically.

Tasks that once required a platoon of specialists - colourists, animators, junior copywriters - can in prototype form be managed by a single creator with a laptop. This isn't to say that those professions will vanish; rather, their role moves up the value chain toward supervision, curation, refinement. The colourist becomes a colour director, setting the emotional palette that AI executes. The animator becomes a movement choreographer. The junior copywriter becomes a brand voice strategist. Meanwhile, the bottom of the pyramid opens to tens of millions of aspirants who were previously locked out by capital constraints.

But here's the paradox of high RODE: when everyone can manifest their dreams with minimal effort, the landscape transforms entirely. Easy access to these tools doesn't guarantee that anyone will notice what you make. When it costs almost nothing to produce music, video or writing, the real bottleneck shifts from creation to attention.

A film you "shoot" in your head with AI will land in a marketplace already filled with thousands of similar experiments. The democratisation of creation leads inevitably to the aristocracy of attention. What still stands out is human taste and honesty, those small, personal signals that software cannot mass-produce - the catch in a voice that speaks of real loss, the frame that lingers just long enough to let meaning settle, the word choice that could only come from lived experience.

This is the RODE paradox: as the ratio approaches infinity (minimal effort, maximum creative output), new forms of scarcity emerge. Not scarcity of tools or techniques, but scarcity of authentic perspective, curatorial brilliance, and the ability to create genuine emotional connection.

Ironically, the easier creation becomes, the more valuable a genuine voice becomes. In a world of infinite content, the human touch - messy, imperfect, but real - becomes the ultimate differentiator. Is the ability to command attention, to harness the implicit intuition of a human, perhaps the truest definition of creativity?

When anyone can make, maybe creativity lies not in the making but in making others care.

The Creative Underclass: Navigating the Transition

While we celebrate the soaring RODE, we must acknowledge those caught in its downdraft. For every creator liberated by AI tools, there's another watching their livelihood evaporate. The revolution isn't painless.

The disruption is real and measurable. Stock photography platforms report fundamental shifts in contributor economics. Getty Images noted in their *2024* investor report that while AI-generated content submissions increased exponentially, traditional photography licensing revenue declined. Translation industry data from CSA Research shows that basic document translation rates have fallen by up to *50%* in some language pairs since *2022*, while demand for "machine translation post-editing" has surged.

New roles are emerging from this disruption. Major agencies including Ogilvy and Wieden+Kennedy have created positions like "Creative Technologist" and "AI Creative Director" - roles that didn't exist three years ago. These positions require both traditional creative skills and fluency in AI tools. The reskilling is happening in real time: online learning platform Domestika reported that their "AI for Creatives" courses became their fastest-growing category in *2024*, with over *2* million enrolments.

The pattern is consistent across creative fields. In publishing, developmental editors who once line-edited manuscripts now focus on story architecture and character development, leaving grammatical polish to AI tools. In advertising, copywriters spend less time generating tagline variations and more time developing brand strategy. In graphic design, professionals report pivoting from pure execution to creative direction and client consultation.

Not everyone successfully navigates this transition. Industry surveys consistently show that mid-career professionals face the greatest challenges - they have deep expertise in methods that are being automated but may lack the flexibility or resources to retrain. Early-career creatives, paradoxically, sometimes adapt more easily, having entered the field with AI as a given rather than a disruption.

We shouldn't pretend disruption isn't happening. It's best to recognize that as AI handles the mechanical aspects of creative work, human judgment, taste, and cultural understanding become more, not less, valuable. The question for every creative professional isn't "Will AI replace me?" but "How can I move up the value chain?"

The Emancipation Beyond the Elite

There's something else that's important when discussing the exciting prospect of 8 billion creatives being liberated.

When I speak of AI democratising creativity, I must acknowledge that "democratisation" means little if the tools remain beyond reach. A grandmother in rural Bangladesh with stories to tell but no internet connection, a dyslexic teenager whose brilliant ideas get trapped behind the written word, a single parent working three jobs with no time or money for creative pursuits – these are the voices that most need amplifying.

The good news is that AI is becoming radically more accessible. Mobile phones – now ubiquitous even in developing regions – can run increasingly sophisticated AI applications. Voice interfaces bypass literacy barriers. Free tiers of AI services lower financial barriers. But we must ensure this trend continues.

For neurodiverse creators, AI can be genuinely transformative. Those with dyslexia can speak their stories rather than write them. People with ADHD can use AI to help structure their scattered brilliance into coherent narratives. Autistic creators might find in AI a patient collaborator that doesn't judge their different ways of processing the world. The artist with motor disabilities can create visual art through text descriptions rather than physical manipulation of tools.

The digital divide remains real. Internet access, device ownership, and digital literacy are not universal. The promise of all these newly-liberated creative voices on Earth will remain hollow unless The Panthropic is universally available. This isn't just about technology – it's about justice.

Listening to the Duet

Practical Steps Forward

So, what should creators and policymakers do with all of this?

First, accept that the technology is already here and cannot be wished away. Trying to ban or ignore it will only leave you behind. The European Union's *AI Act*, the UK's pro-innovation approach, California's SB *1047* debates - all grapple with regulation, but none propose 'uninventing' what exists.

Second, learn the basics of how these systems work - how to write clear prompts, how to spot bias, and how to trace where the training data came from. Stanford's Human-Centered AI institute offers free courses; companies like Anthropic publish research on AI safety; even YouTube tutorials teach prompt engineering. These skills now belong alongside spelling, maths, and colour theory on any curriculum.

Third, remember that people still set the agenda. Even the most independent-looking models run on goals, data, and budgets that humans decide. OpenAI's board governs ChatGPT's values; Google's teams shape Gemini's capabilities; open-source communities guide Stable Diffusion's evolution. The machine can explore, but only we can choose the direction and judge the outcome.

The Invitation

The best habit now is simple curiosity.
Plus engagement.
Get going.
Learn.
Do.

Use AI to help you with recipes at the last moment, with game-plans you're conceiving, with love-letters you're honing, with budget-balancing you're struggling with, with late-night musings about whether your career is worth the effort you're putting in.

Start talking to the Panthropic, to the whole of life and civilisation. It's called AI, but it's also the amalgamation of all us, ready to help you. If we get it right, Panthropia could await us – a world in which borders are culturally porous, where compassion from understanding flows freely, where truth and trust pervade, rather than suspicion and confrontation. I know I push a little too hard here, but when one sees the light through a crack, it's hard not to look at it and wonder what lies outside. Could we be haltingly entering the Panthropocene era, during which humans finally all get along and pull in the same direction?

We stand, all of us, in a similar playground, half-dazzled, half-unnerved.

We can quake at the immensity of the change, or we can quiver - that is, let the shock travel through us like a resonant string - and answer with fresh music, delighted to be creatively enabled.

This book, this chapter, is my invitation to choose the latter.

Chapter 7

The Creative Classroom

From Retention to Revelation: How Education Must Dance

We've witnessed the creative revolution in studios, laboratories, and marketplaces around the world. We've seen artists, scientists, and entrepreneurs learning this new dance with AI. But there's one arena where the stakes are perhaps highest of all - the classroom. If we're truly entering an age of Collaborative Creativity, then we must ask: how do we prepare the next generation for a world where creativity isn't a luxury but a necessity? How do we teach children not just to use AI, but to dance with it while keeping their human spark burning bright?

There is a seismic shift happening in education. We're moving from a model built on retention - memorise, recall, repeat - to one that must emphasise streaming, synthesis, and critical thinking. If creativity is, as I've argued throughout this book, the practice of transforming novel connections into meaningful realities, then education must fundamentally reimagine its purpose.

The old model made sense in an information-scarce world. When books were rare and expensive, when knowledge was locked in distant libraries, the human mind needed to be a storage device. We drilled multiplication tables, memorised poetry, retained dates and facts. The student who could recall the most was deemed the brightest.

But what happens when every student has access to the accumulated knowledge of humanity in their pocket? When AI can retrieve any fact, solve any equation, translate any text in milliseconds? The answer isn't to abandon education but to transform it into something far more powerful: an incubator for creative thinking.

The Shift from Storage to Synthesis

Sir Ken Robinson, who spent his career advocating for creativity in education, argued that "we are educating people out of their creative capacities." In his famous 2006 TED talk, he noted that children who enter school brimming with creative confidence gradually learn to fear being wrong. By the time they graduate, many have lost touch with their creative selves entirely.

This observation becomes even more urgent in the age of AI. As Robinson noted, "If you're not prepared to be wrong, you'll never come up with anything original." Yet our education systems have been designed to reward correct answers over original thinking. In a world where AI can provide all the correct answers, this model becomes not just outdated but actively harmful.

Sugata Mitra, whose "Hole in the Wall" experiments demonstrated that children can teach themselves complex concepts through curiosity and collaboration, offers a different vision. His work showed that when children in Indian slums were given computers with no instruction, they taught themselves not just to use the technology, but to learn English, explore advanced topics, and solve problems collaboratively.

"The teacher's role," Mitra argues, "is not to provide answers but to pose interesting questions." In the AI age, this becomes even more crucial. The teacher becomes a curator of curiosity, a designer of challenges that require creative synthesis rather than rote recall.

The New Creative Curriculum

What would education look like if we designed it specifically to amplify human creativity in partnership with AI? Several pioneering educators are already showing the way.

Finland, celebrated for its world-class education system, has integrated **phenomenon-based learning** into its national curriculum. Rather than completely replacing traditional subjects, schools now dedicate time to interdisciplinary modules where students tackle real-world topics, such as climate change. In these projects, students might apply knowledge from physics, biology, and economics to solve complex problems, often using digital tools and emerging AI to help them research and make connections across different fields.

Progressive approaches to design education exemplify this transformation. These approaches to teaching design don't start with software skills or historical knowledge but with fundamental questions about perception, meaning, and communication. Students learn to see before they learn to make, to question before they learn to answer.

In one exercise, students are asked to document the same object - a chair, perhaps - from multiple perspectives: functional, emotional, cultural, historical. They then use AI to explore how different cultures have solved the problem of sitting. The goal isn't to memorise chair designs but to understand how human needs, cultural values, and material possibilities intersect in the creative act.

Assessment in the Age of AI

If students can use AI to write essays, solve problems, and create presentations, how do we assess learning? The answer, paradoxically, is to make AI use not just allowed but required, and then to assess something deeper than the output.

Sal Khan, founder of Khan Academy, has been exploring this frontier. His team has developed AI tutors that don't give answers but ask amplifying questions, guiding students toward understanding. The assessment isn't whether students get the right answer but how they navigate the problem-solving process.

"We need to move from assessment of learning to assessment for learning", says Dylan Wiliam (a highly influential British educationalist and emeritus professor of educational assessment at the UCL Institute of Education), whose work on formative assessment has influenced educators worldwide. In practice, this means evaluating not what students produce but how they think, how they use tools (including AI), and how they synthesise information into new understanding.

Some practical approaches emerging include:

Process Portfolios: Students document their creative journey, including false starts, iterations, and reflections. AI use is transparent and annotated - students explain why they made certain queries, how they evaluated AI suggestions, and what they learned from the collaboration.

Live Problem-Solving: Students tackle novel challenges in real-time, with full access to AI tools. Assessment focuses on their approach, creativity, and ability to synthesise AI input with human insight.

Creative Partnerships: Students work on projects that require both human and AI contributions, clearly delineating and reflecting on each partner's role. This mirrors the real creative collaborations they'll engage in beyond school.

The Democratisation of Creative Confidence

Perhaps the most profound shift is in who gets to be creative. Traditional education has often identified and nurtured a "creative class", those with apparent talent in the arts, while steering others toward more "practical" pursuits. AI partnership changes this fundamental equation.

Howard Gardner's theory of multiple intelligences suggested that creativity manifests in many forms: linguistic, logical-mathematical, spatial, musical, bodily-kinesthetic, interpersonal, intrapersonal, and naturalistic. AI tools can now amplify each of these intelligences, allowing students to express creativity in their own unique ways.

A student with strong logical-mathematical intelligence but limited drawing skills can now create visual art through AI collaboration. A student with deep interpersonal intelligence can design interactive experiences without coding expertise. The barriers between vision and execution dissolve.

Yong Zhao, a highly-regarded Chinese-American scholar who has written extensively about cultivating creativity in education, argues that, "we need to shift from deficit-based to strength-based education." Instead of focusing on what students can't do, we should amplify what they can do uniquely well. AI becomes a bridge between individual strengths and creative expression.

Case Studies in Creative Education

The **Hasso Plattner Institute of Design, commonly known as the Stanford d.school,** has pioneered "design thinking" in education, teaching students to approach problems with empathy, ideation, and experimentation. They've now integrated AI as a creative partner throughout the process. Students use AI to research user needs, generate solution possibilities, and prototype rapidly. But the emphasis remains on human judgment, ethics, and meaning-making.

The African Leadership University in Rwanda takes a different approach. Recognising that their students will work in a world where AI handles routine tasks, they've eliminated traditional majors in favour of "missions". These are complex challenges that require interdisciplinary thinking. Students learn to use AI as a research and creation tool while developing the uniquely human skills of leadership, ethics, and cultural navigation.

Teaching the Teachers

The transformation of education requires the transformation of educators. Many teachers trained in the retention model feel threatened by AI, seeing it as a force that makes their expertise obsolete.

Linda Darling-Hammond, whose research has shaped teacher education globally, argues for a different perspective: "Teachers need to become learning designers." This means understanding how AI tools work, certainly, but more importantly, understanding how to create learning experiences that leverage both human and artificial intelligence.

Professional development programmes are emerging that help teachers make this transition: learning to use AI as a teaching assistant that can provide personalised feedback at scale; designing assignments that require creative synthesis rather than information retrieval; understanding how to spot and develop different forms of creative intelligence; and creating classroom cultures that celebrate productive failure and iterative improvement.

Initiatives from institutions like **Stanford's d.school** and organisations like **ISTE** train educators to design assignments that require human creativity beyond AI and to build classroom cultures that celebrate iterative improvement. At the same time, tools like **Khanmigo** provide hands-on experience using AI as a teaching assistant for personalised feedback, while frameworks based on the work of **Howard Gardner** help teachers spot and develop the diverse creative intelligences of every student.

It is these programmes that will harness the creativity of future generations.

The Risk of New Inequalities

As with any technological shift, there's a risk that AI could exacerbate educational inequalities rather than reduce them. Students with access to advanced AI tools and guidance on how to use them creatively may leap ahead, while others are left behind.

This is why such pedagogical approaches are so crucial. By focusing on fundamental creative capabilities - seeing, questioning, connecting - rather than specific tools, we can ensure that all students develop the core competencies needed for creative collaboration with AI.

Moreover, as AI tools become more accessible and affordable, there's an opportunity to democratise creative education in unprecedented ways. A student in rural Kenya can access the same AI creative partners as one in Silicon Valley. The challenge is ensuring they also have the educational framework to use these tools meaningfully.

A New Vision for Human Potential

What excites me most about this educational transformation is how it aligns with the creative emancipation I've described throughout this book. When education shifts from retention to creative thinking, when AI handles the drudgery of recall and calculation, human potential is unleashed in new ways.

Students no longer need to spend years mastering technical craft before they can express creative vision. They can engage with big ideas, complex problems, and creative expression from the beginning. The gradient of learning becomes less steep, the joy of creation more immediate. What seemed precocious before now, might seem laggardly today. The world is accelerating.

This doesn't mean abandoning rigour or depth. True creativity requires deep understanding, critical thinking, and reflective practice. But these qualities develop through creative engagement, not despite it.

Conclusion: The Classroom as Creative Laboratory

The transformation of education from retention to revelation isn't just a pedagogical shift, it's a fundamental reimagining of human potential. When every student has access, we're not just changing how we teach. We're changing who gets to create.

The universal creative pulses I've celebrated throughout this book include millions of young minds currently sitting in classrooms. Whether those creative sparks are nurtured or extinguished depends on how quickly and thoughtfully we can transform education for the AI age.

In this new model, every classroom becomes a creative laboratory, every student a potential innovator, every teacher a guide to the possible. The AI that might have replaced human thinking becomes the partner that amplifies it and yields to its power to feel.

The dance between human and machine isn't just reshaping our creative industries. It's revolutionising how we develop human potential itself. And in that revolution lies perhaps the greatest creative act of all: reimagining what it means to learn, to grow, and to become a more fully realised human in the age of AI.

Chapter 8

Nations in the dance

How Countries are Choreographing their Creative Futures

Individual creators can embrace AI tomorrow. Companies can transform their workflows next quarter. But what happens when entire nations recognise that creative collaboration with AI isn't just another technological shift, but a fundamental reimagining of human potential?

Around the world, governments are awakening to a profound realisation: in the 21st Century, a nation's creative capacity may matter more than its natural resources. The countries that learn to choreograph this dance between human imagination and machine intelligence won't just compete - they'll help define what it means to be human in the age of AI. Let's see how different cultures are approaching this civilisational moment.

In *2017*, the **United Arab Emirates (UAE)** did something unprecedented. They appointed Omar Al Olama as the world's first Minister of State for Artificial Intelligence. He was *27* years old. The message was clear: AI wasn't just another technology to regulate; it was a transformation requiring cabinet-level leadership.

Seven years later, the global landscape reveals a fascinating patchwork of national strategies, each reflecting different cultural values, economic priorities, and creative ambitions. Some nations are orchestrating comprehensive transformations; others are taking tentative first steps. What emerges is a picture of humanity collectively grappling with how to harness AI not just for efficiency or security, but for creative efflorescence.

The Pioneer Nations: Comprehensive Strategies

The **UAE's** appointment of an AI Minister was just the beginning. The country has since launched its National AI Strategy *2031*, explicitly linking AI development to creative industries. The strategy includes establishing AI creativity labs in universities, funding AI-assisted film production, and creating regulatory sandboxes where artists can experiment with AI tools without legal uncertainty. Dubai's Museum of the Future showcases AI-generated art and architecture, positioning the emirate as a hub for tech-enabled creativity.

Singapore's National AI Strategy takes a different but equally comprehensive approach. Their focus on becoming a "Smart Nation" includes specific provisions for creative industries. The Infocomm Media Development Authority provides grants for AI-creative projects, from interactive digital art installations to AI-assisted music composition. Their "AI for Everyone" programme ensures that creative professionals, not just technologists, understand and can leverage these tools.

China's approach is perhaps the most ambitious in scale. The "New Generation Artificial Intelligence Development Plan" explicitly identifies creative industries as a key application area. Chinese schools are mandating AI education from primary level, including creative applications like AI-assisted storytelling and digital art. Companies like Baidu and Alibaba are creating AI tools specifically designed for Chinese creative traditions, from calligraphy to classical music composition.

Finland takes a uniquely democratic approach, exemplified by its "Elements of AI" course, which is available free to all citizens. The initial goal was for 1% of the population to gain AI literacy, a target chosen strategically to create a critical mass of people who could innovate and spread their understanding throughout society, giving the nation a competitive advantage without trying to make everyone a programmer. This core strategy is supported by special programmes for artists and designers and by funding for cultural institutions to experiment with AI in fields from game design to architecture.

The Creative Focus: Nations Prioritising Cultural Innovation

South Korea stands out for its explicit focus on AI in creative industries. The Ministry of Culture, Sports and Tourism launched "Culture+AI" initiatives that fund K-pop producers experimenting with AI composition, filmmakers using AI for visual effects, and game developers creating AI-driven narratives. Given Korea's global cultural influence through the Hallyu wave, this positions them to export AI-enhanced creativity worldwide.

France's "AI for Humanity" strategy includes significant provisions for cultural preservation and innovation. The Bibliothèque Nationale de France uses AI to digitise and make searchable centuries of French literature and art. French film subsidies now include provisions for AI-assisted production, while maintaining requirements for human creative control. It's a balance between embracing innovation and protecting cultural authenticity.

Japan's Society 5.0 vision integrates AI throughout society, with particular emphasis on creative applications that respect Japanese aesthetics and values. From AI that can generate haiku following traditional rules to systems that assist in manga production while preserving artistic style, Japan seeks to enhance rather than replace its creative traditions.

Canada's Pan-Canadian AI Strategy includes significant investment in creative applications through organisations like the Canadian Institute for Advanced Research (CIFAR). The National Film Board of Canada experiments with AI-driven interactive documentaries, while the Canada Council for the Arts funds artists exploring AI collaboration.

The Education Revolutionaries

Several nations are reimagining education specifically around AI and creativity. **Estonia**, already a digital pioneer, has introduced AI and robotics education from age 7, with creative problem-solving at its core. Students learn not just to use AI tools but to think creatively about their applications.

Rwanda's partnership with the African Institute for Mathematical Sciences includes AI education focused on solving African challenges creatively. Students use AI to design solutions for agriculture, healthcare, and urban planning, blending technical skills with creative thinking.

The Measured Approaches

Not all nations are moving at the same pace, and some deliberately so. **Germany's** AI strategy emphasises ethical considerations and worker protections, including for creative professionals. Their approach seeks to ensure AI enhances rather than replaces human creativity, with strong unions advocating for creative workers' rights.

The **Nordic countries** (beyond Finland) take a characteristically balanced approach. Denmark's "National Strategy for Artificial Intelligence" includes provisions for creative industries but emphasises human agency. Sweden funds research into AI's impact on creative work, ensuring evidence-based policy. Norway uses its sovereign wealth fund to invest in AI education that maintains their tradition of democratic participation.

The **United Kingdom's** National AI Strategy includes creative industries as a key sector, but implementation has been gradual. The Creative Industries Council works with government to ensure AI adoption doesn't undermine the UK's creative economy.

Initiatives like the Creative Industries Policy and Evidence Centre research how AI can enhance rather than threaten creative work.

Despite being home to leading AI companies like OpenAI, Google, and Meta, the **United States** ranks surprisingly low in global AI adoption indices - often outside the top *10* in implementation metrics. While the US leads in AI research, development, and venture capital investment, it lags behind countries like Singapore, Denmark, and South Korea in actual deployment across industries and government services. This gap between innovation and implementation reflects regulatory uncertainty, legacy system entrenchment, and a more fragmented approach compared to countries with national AI strategies. The US excels at creating breakthrough AI technologies but struggles with the systematic integration that smaller, more coordinated nations achieve.

The world, with AI, is changing.

Grassroots Innovation: When Communities Lead

Not every country's AI journey is led from the top down. New Zealand's experience reveals the power of grassroots innovation in the absence of national coordination. Despite being the last OECD country to release an AI strategy in *2024*, with education conspicuously omitted, remarkable initiatives have emerged from the ground up.

Te Hiku Media, a Māori radio station in Kaitaia, has built one of the world's most sophisticated community-owned AI systems for indigenous language preservation. Using an advanced computer purchased at a discount and housed in what they describe as "a derelict rural town", they've created AI models for te reo Māori built exclusively from consented community data. Within ten days of launching a community challenge, they amassed *310* hours of speech from *2,500* people. Money wasn't the only motivator. People trusted Te Hiku to act as kaitiaki (guardians) of their data.

Individual teachers, too, are experimenting with AI to enhance Māori storytelling and mātauranga (knowledge), even without official curriculum guidance. This creates a fascinating tension: vibrant innovation happening despite, rather than because of, national policy.

The lesson? Sometimes the most important AI developments don't wait for government strategies. It also raises questions about equity and scale. What works in a passionate community might not reach those without the same resources or drive. New Zealand's story suggests that whilst grassroots innovation is powerful, it cannot replace the systematic support that national strategies provide.

The Challenges Ahead

This global adoption isn't without its tensions. Nations must balance several competing priorities:

Cultural Authenticity vs Global Competitiveness: Countries with strong creative traditions worry about AI homogenising their unique cultural expressions. The challenge is using AI to amplify rather than dilute cultural identity.

Economic Opportunity vs Worker Protection: While AI can democratise creativity, it also threatens traditional creative jobs. Nations must navigate between embracing efficiency and protecting livelihoods.

Access vs Excellence: Should countries focus on giving everyone basic AI creative tools or on developing cutting-edge capabilities for specialists? Different nations are making different choices.

Regulation vs Innovation: Too much regulation stifles creative experimentation; too little risks ethical problems and cultural damage. Finding the balance is proving difficult globally and raises fundamental questions about who should govern a technology that transcends borders.

Should there be a global "RegulAItor"? The idea has appeal - a unified body setting standards for training data, ethical use, and creative rights. But whose values would it embody? What works for Silicon Valley might stifle innovation in Shenzhen or feel culturally tone-deaf in São Paulo. The internet never received a global governor, and perhaps that's why it flourished.

National approaches reflect cultural priorities. The EU's AI Act emphasises citizen protection and rights. China focuses on social stability and industrial leadership. The US relies more on market forces and voluntary guidelines. The UK positions itself as the sensible middle ground. Each approach has merit; none is complete.

Perhaps the answer isn't a single regulator but a coalition of standards bodies - like how the internet evolved through consortiums such as W3C. An "AI Standards Agency" could establish technical benchmarks without imposing cultural values. Think ISO standards for AI: not exciting but effective.

The wild card is whether AIs could self-regulate. Not in some science-fiction sense of machines governing themselves, but through technical architectures that build in ethical constraints. Models that refuse to generate harmful content, that watermark their creations, that track provenance automatically. The regulation embedded in the code itself.

Most likely, we'll muddle through with a patchwork - some international agreements on the worst abuses, national laws reflecting local values, industry self-regulation where it works, and technical standards that become de facto rules.

What Success Looks Like

The most successful national approaches share certain characteristics:

Leadership Recognition: Whether through ministers, strategies, or significant funding, successful nations treat AI creativity as a priority.

Educational Integration: Countries seeing real transformation are teaching AI creativity from early ages, not just to specialists.

Cultural Sensitivity: The best strategies enhance local creative traditions rather than replacing them with global defaults.

Democratic Access: Nations creating broad-based creative revival ensure AI tools reach all citizens, not just elites.

Ethical Frameworks: Successful adoption includes protections for creative workers and respect for intellectual property.

The Global Creative Renaissance

What strikes me most about this global adoption is its diversity. There's no single model being imposed worldwide. Instead, each nation is finding its own dance with AI, reflecting its values, strengths, and creative traditions.

This diversity drives optimism. Rather than AI creating a homogenised global culture, we're seeing it amplify the unique creative voice of each nation. Korean AI doesn't look like Finnish AI; Emirati applications differ from Japanese ones. The technology is universal, but its creative application remains beautifully, necessarily local.

We're witnessing something unprecedented: a coordinated yet diverse global effort to enhance human creativity through technology. Not every nation will succeed equally, and there will be setbacks and corrections. But the direction is clear. Countries worldwide recognise that in the age of AI, creativity isn't just a nice-to-have; it's essential for economic competitiveness, cultural vitality, and human flourishing.

The billions of creative pulses I've celebrated throughout this book aren't just individual heartbeats. They're organised into nations, each finding its own rhythm in the dance with AI. And in that global choreography lies perhaps our greatest hope: that AI will not flatten human creativity into sameness, but help each culture sing its unique song more powerfully than ever before.

Chapter 9

Constructive Contradiction

A Dialogue with those who Oppose AI's Role in Creativity

To champion a new way of seeing the world without first listening, truly listening, to those who view it with suspicion would be an act of arrogance. I listen to those who are sceptical about AI and especially its role in creativity. An argument is only as strong as its ability to withstand the most intelligent critiques levelled against it.

Some people truly believe this technology is no more than a hyped-up, money-spinning machine built on stolen, copyrighted assets. These are smart, informed, careful people, not knee-jerkers who feel threatened and bring protectionism to bear immediately upon their own endeavours. They deserve to be heard.

As a psychologist, I recognise their caution as a natural response to imminent change. Cognitive dissonance pushes us first toward defence, then, if handled with respect, toward dialogue.

Throughout this book, I've laid out an optimistic case for a new creative alliance, a duet between human intuition and machine intelligence. It's a vision of emancipation, of new voices, and of a more expansive creative future.

But this isn't the only story being told.

There are other, equally or more intelligent and passionate voices that raise profound and necessary objections. These aren't the reactions of Luddites, but the considered arguments of artists, philosophers, and computer scientists who look at this new technology and see not a partner, but a parasite; not a tool for liberation, but a machine for homogenisation; not a new form of creativity, but the devaluation of what it means to be human.

The most ardent proponents of a new technology can sometimes be the least interesting. I don't want to be one of those. True understanding comes from a dialogue with doubt. In the spirit of honest inquiry, this chapter is dedicated to the doubters, the sceptics, the rejectors. This chapter is a record of what I call constructive contradiction.

These are the voices of the loyal opposition, and their role is to test our assumptions, to challenge our optimism, and to force us to confront the uncomfortable truths that lie beneath the surface of the hype. Their contradictions aren't an obstacle to our journey; they're an essential part of it.

This isn't a cage fight; it's a true discourse, a real discussion, and I'm grateful for their independence of thought and stirring, contrary views.

Let's listen to them.

Part *1*: The Intellectual's Critique - The Question of "Mind"

The first and most fundamental set of objections comes from thinkers who challenge the very nature of AI's intelligence. They argue that we're making a profound category error when we attribute understanding, reasoning, or creativity to these systems. Their critique isn't about whether the output is useful, but whether there's any genuine "thought" behind it.

The Stochastic Parrot and the Grounding Problem

One of the most potent and widely cited critiques comes from the computational linguistics professor **Emily M. Bender** at the University of Washington. In her influential *2021* paper "On the Dangers of Stochastic Parrots" (co-authored with Timnit Gebru, Angelina McMillan-Major, and Margaret Mitchell), she articulated the argument that LLMs are best understood not as nascent minds, but as "stochastic parrots". I'm reading her excellent and challenging book now, called '*AI Con*' in which she develops these arguments.

The metaphor is precise. A parrot can mimic human speech. It can string together complex phrases and might even appear to respond. But the parrot has no understanding of the meaning behind the words it speaks. It's repeating statistical patterns it's heard, without any connection to the real-world concepts those words represent.

Bender argues that LLMs operate in the same way. They've been trained on a colossal corpus of text and can therefore predict, with impressive accuracy, which word is most likely to follow another in any given context. They can generate grammatically correct, stylistically coherent, and often plausible sentences.

But, she contends, this is mimicry, not meaning. The AI doesn't "know" what a cat is. It's never felt the softness of its fur or heard its purr. It only knows that the word "cat" frequently appears in its training data alongside words like "meow," "whiskers," and "sofa".

Bender illustrates this with a brilliant thought experiment. Imagine an octopus, isolated in a deep-sea cavern, that manages to tap into a transatlantic internet cable. It can read all of humanity's text, learning every pattern and correlation. It could, in theory, learn to communicate with a human on the other end of the line so convincingly that it could easily pass the Turing Test. But could the octopus ever truly understand what "a warm summer's day" means? It has no body, no senses, no shared physical context with which to ground the meaning of those words. It has the text, but it doesn't have the world.

This leads to her second, equally important concept: the grounding problem. For language to have meaning, it must be grounded in lived experience. Words aren't just patterns; they're symbols that point to real things in the world. Without that connection to reality, language is just empty syntax.

From this perspective, our creative duet with AI is a dance with an illusionist. The machine's output feels meaningful because we project meaning onto it. We fill in the gaps. We see a coherent sentence and assume a coherent mind behind it. For Bender, the throne is empty. Any creativity we perceive is a projection of our own desire to find a mind in the machine.

Takeaway: fluency isn't understanding; language needs grounding, not just patterning.

The Kludge and the Limits of Brute Force

A complementary critique comes from cognitive scientist **Gary Marcus** at New York University. While Bender focuses on meaning, Marcus focuses on engineering reality. In his book "*Rebooting AI*" and numerous articles, he calls current systems a "kludge": clumsy, temporary fixes that work until they don't.

Marcus points to brittleness. Models can write a sonnet but fail at simple arithmetic. They can draft a legal brief but invent case law. They lack what he calls "deep understanding" and common-sense reasoning. Their success is a function of brute force: train on enormous data, and many problems yield through statistical pattern-matching. But when faced with a situation outside those patterns, they fail, often unpredictably.

For Marcus, the idea that scale alone will deliver general intelligence is an illusion. He believes progress now requires a different architecture: innate knowledge, symbolic reasoning, structured causality. Current AI is like a student who's memorised the textbook but doesn't understand the principles.

From a creative standpoint, Marcus warns against mistaking fluency for insight. An AI can generate a plausible story or image, but it cannot reason about cause and effect, character motivation, or the nuances of human relationships. Its creativity is surface-level recombination of training data.

In a *2023* debate with Yann LeCun at NYU, Marcus demonstrated this by asking an AI simple questions about physical reality: "If I put cheese in the fridge, will it melt?" The AI's inconsistent answers revealed its lack of genuine world knowledge. If it can't understand basic physics, how can it truly create?

Takeaway: scale brings reach, not depth; brute force cannot replace reasoning.

Data Colonialism and Cultural Flattening

Computer scientist **Timnit Gebru**, now leading the Distributed AI Research Institute, widens the lens to include power and geography. She argues that large AI models rely on data extracted, often without consent, from communities that rarely share in the resulting value. She calls this "data colonialism": a continuation of historical patterns where resources flow from the many to the few.

In her research, Gebru has shown how AI models trained primarily on English-language text from wealthy nations perpetuate specific worldviews. When she analysed image generation models, she found they associate "beauty" with European features, "poverty" with African settings, and "technology" with East Asian or Western contexts. These aren't neutral tools but systems that embed and amplify existing power structures.

The creative risk is cultural flattening: local voices drown as dominant languages and styles saturate the training pool. When an AI is trained primarily on Hollywood films, Western literature, and English-language web content, it perpetuates those perspectives while marginalising others. For every voice amplified, how many are silenced?

Gebru's work forces us to ask uncomfortable questions: Whose creativity is being emancipated? At what cost to others? The panthropic vision I've championed assumes equal representation, but the reality is far more complex and inequitable.

Takeaway: those who train the model shape which voices are amplified, and which are erased.

The High-Tech Plagiarist

Computer scientist and virtual reality pioneer **Jaron Lanier** frames the critique in economic terms. In his book "*Who Owns the Future?*" and recent essays, he argues that generative AI models are engines of "high-tech plagiarism".

These models don't create anything genuinely new, he argues. Their output is an untraceable remix of the work of millions of human creators whose data trained them. When an AI generates an image "in the style of" a particular artist, it performs a mathematical operation that appropriates that artist's discoveries without consent, credit, or compensation.

Lanier sees this as a fundamental devaluation of human creativity. Original creators are obscured while profits flow to the corporations that own the models. He warns of a culture of "content farming", where the goal is to generate endless streams of derivative material that erodes the information ecosystem.

A recent example: in *2024 The New York Times* sued OpenAI and Microsoft, alleging billions in damages for unlicensed use of newsroom text. The legal fight underscores Lanier's point about value extraction. Similar suits from authors (including John Grisham and George R.R. Martin) and visual artists (in class actions against Stability AI and Midjourney) show this isn't theoretical.

Takeaway: if compensation detaches from creation, the creative economy corrodes.

Part 2: The Artist's Rejection - The Question of "Soul"

If the intellectuals question the "mind" of the machine, the artists question its "soul". Their rejection isn't based on a technical analysis of how the models work, but on a lived understanding of where art comes from. For them, creativity is a human process, rooted in the messy, painful, and necessary business of being alive.

The Absence of Suffering: Nick Cave

My publisher, Richard Charkin, loves Nick Cave. So do I, especially his *Red Hand Files*, where he fields questions every week with empathy and candour. He's a living, interactive, creative force.

No one has articulated the "absence of suffering" argument more clearly. When a fan submitted a ChatGPT song "in the style of Nick Cave", Cave responded with a public letter in January 2023. He called the result a "grotesque mockery". His objection was existential: art is born from struggle, "the internal, creative struggle", as he puts it. An AI has no inner being, has been nowhere, has endured nothing. It's never had a broken heart or stood at a graveside. Therefore, its work, to Cave, is a technically competent but soulless imitation.

Cave's argument cuts to the heart of what many artists feel: that creativity isn't just about arranging words or colours in pleasing patterns. It's about transmuting lived experience into form. The AI has no experience to transmute. It has data, but not life.

This isn't mere romanticism. Cave points to something fundamental about human creativity: it emerges from our mortality, our suffering, our joy. These aren't bugs in the human system; they're features. They're what give art its weight, its resonance, its ability to move us.

In a follow-up Red Hand File, Cave wrote: "It could perhaps in time create a song that is indistinguishable from an original, but it will always be a replication, a kind of burlesque... Songs arise out of suffering, by which I mean they are predicated upon the complex, internal human struggle of creation and, well, as far as I know, algorithms don't feel."

Takeaway: authentic art costs something; without lived experience, output rings hollow.

The Sanctity of the Mess: Justine Bateman

American filmmaker, author, and former actor Justine Bateman has become one of AI's most articulate critics. In her book *"Face: One Square Foot of Skin"* and numerous interviews, she focuses on process and consent. She argues that attempts to "optimise" filmmaking with AI misunderstand what makes art interesting.

In a *2023* SAG-AFTRA panel, Bateman explained: "Tech aims to remove friction, mistakes, and accidents. But this 'mess' is the very source of creativity. The unexpected line readings, the logistical constraints, the happy accidents - these give a work its soul." Paraphrasing, she says that craft is part of art.

Her second argument concerns consent. Using an actor's likeness or a writer's work to train AI without explicit permission is, to her, a fundamental violation. An actor's face and voice aren't data points; they're expressions of identity and craft. She's particularly concerned about deceased performers being digitally resurrected without their consent, calling it "digital necromancy".

Bateman's critique highlights how the efficiency-minded tech world fundamentally misunderstands the creative process. Art isn't a problem to be solved or a process to be optimised. It's an exploration, and explorations require getting lost.

During the *2023* Hollywood strikes, Bateman's arguments helped shape union demands for AI protections. The resulting agreements require consent for digital replicas and compensation for AI training use - a direct response to her advocacy.

Takeaway: creativity is embodied and consensual; remove either and you flatten the result.

The Brand Builder's Critique: Zoe Scaman

Strategist **Zoe Scaman** brings a different perspective from the world of brand building. In her influential newsletter and talks, she argues that what makes a brand resonate isn't clever copy or beautiful visuals - it's the patient, careful construction of a coherent "world" that has genuine cultural relevance.

In her essay "The Empty Internet," Scaman warns that AI-generated content creates what she calls "cultural static" - technically proficient material that lacks the specific point of view that makes work memorable. She writes: "Brands aren't built on efficiency. They're built on friction - the specific, sometimes difficult choices that create distinctive character."

This kind of deep, strategic work - understanding cultural currents, identifying authentic insights, building meaningful narratives - requires human intuition and cultural fluency that AI cannot replicate. An AI might generate a thousand taglines, but it cannot feel which one will truly resonate with a specific community at a specific cultural moment.

Scaman's critique extends beyond advertising to all forms of cultural creation. The most impactful work doesn't just execute well; it understands and shapes the cultural conversation. This requires not just pattern recognition but genuine cultural participation.

In a *2024* talk at Cannes Lions, she demonstrated this by showing AI-generated campaigns next to human-created ones. The AI work was technically perfect but culturally hollow - "like muzak for the eyes", as she put it.

Takeaway: cultural resonance requires participation, not just observation; AI watches culture but cannot live within it.

Part 3: The Strategist's Scepticism - The Question of "Depth"

Beyond questions of mind and soul, there's a pragmatic critique about the depth and sustainability of AI-assisted creativity. These voices don't necessarily oppose the technology, but they question whether it can deliver on its promises.

The Homogenisation Hypothesis

Media theorist **Douglas Rushkoff** warns of a "grey goo" scenario for culture. In his book *"Survival of the Richest"* and recent talks, he argues that when millions of creators use the same AI models, trained on the same data, we'll see convergence toward a bland mean.

The unique, the challenging, the genuinely novel - these emerge from individual perspectives pushing against convention. But AI, by definition, learns convention. It's trained on what exists, not what could be. Rushkoff demonstrated this at a *2024* conference by showing how AI-generated stories all follow similar narrative structures, regardless of prompt variations.

His concern isn't just aesthetic. A homogenised culture is a weakened culture, less able to adapt, innovate, or challenge power. Diversity isn't just nice to have; it's essential for cultural evolution. "We're creating a feedback loop," he warns, "where AI trained on human culture creates culture that trains the next AI, each generation becoming more recursive and less surprising."

I observe that human stories follow remarkably similar patterns - something Joseph Campbell mapped in "The Hero's Journey" and Christopher Booker identified as "The Seven Basic Plots." From Gilgamesh to Star Wars, from ancient myths to modern movies, we keep telling variations of the same fundamental stories.

This complicates Rushkoff's critique in an interesting way. If humans also gravitate toward conventional narrative structures, what exactly distinguishes human from AI storytelling?

The difference might lie not in the structures but in how we break them. Humans don't just follow patterns - we consciously subvert them. Joyce exploded narrative in *Ulysses*. Beckett stripped story to its bones in *Waiting for Godot*. Charlie Kaufman turned screenplay structure inside out with *Adaptation*. We know the rules precisely so we can break them meaningfully.

AI, trained on the aggregate, tends toward the center of the bell curve. It learns that stories "should" have three acts, that heroes "should" face challenges, that endings "should" provide resolution. It's very good at producing the 50th percentile story. But the stories that matter often live at the edges - too weird, too specific, too rule-breaking to emerge from statistical averaging.

The real question isn't whether AI can learn narrative patterns (clearly it can, just as we do) but whether it can learn when and why to violate them. Can it develop the punk sensibility that says, "everyone does it this way, so I'll do it that way"? Can it capture the specific life experience that makes someone tell a universal story in a completely unprecedented way?

Human convention-following comes from our shared psychology and culture. AI convention-following comes from statistical probability. Both produce patterns, but perhaps only one can produce meaningful rebellion against those patterns.

The Deskilling Dilemma

Education researcher **Sherry Turkle** at MIT raises concerns about what happens to human capabilities when we outsource cognitive work. Her research on calculators and GPS systems shows a pattern: as tools become more capable, humans become less so. Mental arithmetic atrophies; spatial navigation withers.

In her *2023* study of college students using AI for writing, she found alarming patterns. Students who regularly used AI for essay structure showed decreased ability to organise thoughts without it. "They've outsourced not just the labour of writing," she reports, "but the cognitive work of thinking through problems."

Applied to creativity, this suggests a troubling future. If young writers never learn to structure arguments because AI does it for them, if artists never master composition because generative models handle it, what happens to human creative capacity? We risk raising a generation of directors who cannot create, only curate.

Turkle's proposed solution isn't to ban AI but to use it differently - as a teaching tool that makes its reasoning visible, not a black box that simply provides answers. But she worries the economic incentives push toward dependence, not education.

The Attention Economy Argument

Tech critic **Cory Doctorow** points to a fundamental economic problem. In his concept of "enshittification" and recent essays, he argues that when creation becomes frictionless, the real bottleneck isn't making things but getting noticed.

When anyone can generate professional-looking content, the marketplace becomes flooded. But attention remains finite. We're already drowning in content; AI threatens to multiply that problem exponentially. Doctorow calculated that if just *1%* of internet users generate one AI article daily, we'd add *50* million pieces of content every day - more than all newspapers globally produced in the 20th century.

In this view, AI doesn't democratise creativity; it creates a new elite of those who can capture attention in an even more crowded marketplace. The promise of emancipation becomes a reality of even fiercer competition. Small voices don't get amplified; they get drowned out by the exponential increase in noise.

A Dialogue with Doubt: The Path Forward

Where does this leave us?

We've listened to the voices of loyal opposition. Intellectuals challenge the "mind" of AI; artists challenge its "soul"; strategists question its cultural depth and economic impact. These concerns aren't fearful ramblings. They're guardrails as we build.

The critiques from Bender and Marcus push us toward rigorous, reliable AI. The moral arguments from Lanier, Cave, Bateman, and Gebru force us to confront ethics and consent. They demand systems that respect creators and communities. The strategic concerns from Scaman, Rushkoff, Turkle, and Doctorow remind us that technology alone doesn't determine outcomes - human choices do.

And yet... I don't believe these contradictions invalidate the thesis of this book. They refine it. The future they warn against, a world of soulless, derivative, plagiarised content, is possible, but not inevitable. It's a choice.

The path forward, advocated here, accepts that AI isn't a mind like ours. Its strength lies in remarkable analytical execution, not non-existent intuitive feeling. It cannot suffer, cannot truly understand, cannot participate in culture as we do.

But it can still be a partner.

I support a model of Collaborative Creativity where the human provides vision, taste, lived experience, and ethics. The Panthropic connection isn't a replacement for individual perspective, but an unprecedented knowledge tool that can enlarge it. The key is maintaining human agency, human judgment, human soul at the centre of the creative act.

To address the specific concerns raised:

On the stochastic parrot problem: Yes, AI lacks grounded understanding. But in partnership with humans who do have that grounding, it becomes a liberating tool for exploring and articulating human experience. The AI helps us understand our own grounding better.

On data colonialism: This is why I advocate for ethical AI training that compensates creators and includes diverse voices. The solution isn't to abandon AI but to build it better, with equity and representation at its core.

On the absence of soul: Exactly right - AI has no soul. That's why human creativity remains essential. The AI amplifies human soul; it doesn't replace it. Cave's suffering, Bateman's embodiment, Scaman's cultural insight - these remain uniquely human contributions.

On homogenisation: This is a real risk, which is why I emphasise the importance of human taste, curation, and the injection of personal voice. We must actively resist the gravitational pull toward the generic.

On deskilling: We must teach AI collaboration as a skill that enhances rather than replaces human capabilities. Like learning to play an instrument makes you a better listener, learning to work with AI should make you a better thinker.

The thinkers in this chapter have drawn clear lines. They've shown us where the dangers lie. I acknowledge that some go further. There are those who see AI development as 'narcissism at scale,' driven by 'EQ-deficient, dopamine-fuelled' tech elites who 'conflate intelligence with wisdom.' They view the entire enterprise as theft - 'the largest theft in human history' - built on stolen data and environmental destruction."

Our task is to heed their warnings and then consciously walk the other path: toward augmentation not replacement, diversification not homogenisation, and emancipation not exploitation of the artist.

Their contradictions are constructive because they force us to be better. They demand that we build not just effective systems, but ethical ones. They insist that we consider not just what we can do, but what we should do.

This is the value of the loyal opposition. They keep us honest. They prevent us from sleepwalking into a diminished future. They ensure that our quiver of excitement is tempered with appropriate caution.

The dance continues, but now we know where the floor is slippery.

Chapter 10

Becoming a collaborative creator

A Practical Guide to the New Duet

We've travelled a great distance together. We began with the deep, internal psychology of our own creativity, that energetic force with which we're born. We journeyed into the strange, new landscape of an artificial mind, exploring its dynamic but inorganic, and yet allied, intelligence. We've walked through the studios, labs, and workshops of our world, witnessing the practical application of this new partnership. We've paused to listen, with respect, to the constructive contradictions of the loyal opposition, the intelligent and welcome voices of scepticism.

Let me dwell on the 'inorganic' piece for a moment.

There are several fascinating developments suggesting AI might have a more "organic" future than we currently imagine:

Biological Computing: Researchers at Cortical Labs have created "DishBrain" - living brain cells that learned to play Pong. These biological neurons, grown on computer chips, demonstrate that future AI might literally incorporate organic components. Similarly, companies like Koniku are building "wetware" - biological neurons integrated with silicon chips for enhanced computing.

Bio-inspired Architecture: The next generation of AI models increasingly mimics biological systems. Spiking neural networks replicate how biological neurons fire. Neuromorphic chips like Intel's Loihi physically mirror brain structure. These aren't just metaphors - they're attempts to capture the efficiency and adaptability of organic intelligence.

Self-Modifying Systems: While current AI systems have fixed architectures after training, researchers are exploring more adaptive approaches. Some work on "continual learning" models to update with new information without forgetting old knowledge. Meta-learning creates AI that "learns how to learn" more efficiently. However, true self-modification - where AI fundamentally rewrites its own architecture during operation, similar to how brains rewire through neuroplasticity - remains theoretical. Current systems adapt their parameters, not their fundamental structure.

Hybrid Intelligence: Perhaps most intriguingly, Neuralink suggests the boundary between organic and inorganic intelligence might dissolve entirely. If human brains directly interface with AI systems, the distinction becomes meaningless. The intelligence would be neither fully organic nor fully artificial - but something new.

Synthetic Biology: As AI helps design new biological systems, and biological systems inspire new AI architectures, we might see convergence. Imagine AI systems that grow rather than being built, that heal rather than being repaired, that evolve rather than being updated.

The term "dynamic yet inorganic intelligence" might be perfectly accurate for today's AI. But tomorrow's intelligence might be dynamic and increasingly organic - not in origin, but in operation, adaptation, and perhaps even in physical substrate. The dance between human and machine might become a dance between different forms of life itself.

Through our journey together here, a set of core ideas has emerged, a handful of compass points to help us navigate this new territory. Before we take our final step, from understanding to action, it's worth holding these big thoughts in our hands one last time, to feel their shape and weight.

First, we've recognised that the creative process is a duet between two distinct modes of thinking: the intuitive, associative, fast-feeling spark, and the deliberate, analytical, slow-thinking structure. This is the fundamental psychological rhythm of creation.

Second, we've established that the human-AI alliance is a new and remarkable distribution of these roles. We, as humans, provide the spark, the lived experience, the emotional context, the "why". The AI provides the polished, indefatigable analytical execution, the structuring, the iteration, the "how".

Third, we've given a name to the thrilling new state this partnership enables: Panthropia. It's the experience of being in a direct, personal dialogue with the abstracted essence of all human knowledge and expression, a chance to connect with the totality of our shared culture and accomplishments as human beings. AI, one might argue, is the distillation of humanity.

Fourth, we've seen that this isn't a future-tense proposition. This creative collaboration is already reshaping every industry, from the arts to the sciences, from the global to the personal. The new world is already here.

And finally, this has led us to a profound and provocative question about the very nature of human endeavour. In a world where AI can execute our intentions with increasing autonomy, what becomes of the concept of "work" itself? Does this new alliance offer us a path beyond the drudgery of the mill, towards a future where more of us are free to do what humans do best: to feel, to think, to dream, to imagine, and to create?

With these compass points in hand, the final question is a personal one. It's not about what's happening in the world, but about what you, the reader, will do next. This chapter is a practical guide to answering that question. It's a handbook for becoming a collaborative creator.

Part 1: The Mindset - How to Think Like a Collaborative Creator

Before we touch a keyboard, the most important work is internal. The effectiveness of this new partnership isn't determined by the sophistication of the technology, but by the mindset of the human who wields it. To dance this new dance, we must first learn the posture, the attitude, the internal stance of a true collaborator. I believe this starts by leaving aside questions about what AI is, and instead focusing on what we can achieve in creative collaboration with it.

Embrace the Cyborg

The first step is to discard the old model of a master and a servant. As Ethan Mollick describes in his smart and constructive book *"Co-Intelligence"*, the most effective way to work with AI is to adopt what he calls the "cyborg" approach. This isn't about handing off a task to a machine and waiting for the result. It's about creating a tight, fluid, and iterative feedback loop, where the human and the machine are seamlessly integrated.

Think of it as a conversation, not a command or demand. You provide an initial idea. The AI responds. You react to its response, refining your idea. The AI incorporates your new thought and responds again. In this model, the task is passed back and forth, sometimes dozens of times, with each participant enhancing the work of the other. It's a process of co-evolution, as Mollick terms it.

Adopting this mindset means seeing the AI not as a vending machine for answers, but as a partner for your thoughts. It means being willing to be surprised, to be challenged, and to have your own ideas changed for the better through the dialogue.

In this iteration and reiteration, the AI will never flag. You, the human creator, will sometimes feel unsettled, sometimes know something's wrong, sometimes decide enough is enough. That gut instinct, that implicit ability to judge the moment, to identify creative completion - keep that in the foreground. It's how we're different from machines. It's our unique capability. It's precious. As

Lao Tzu wrote: "He who knows that enough is enough will always have enough."

Cultivate Calibrated Trust

As we discussed in Chapter 2, learning to trust a non-human intelligence isn't intuitive. It requires a new psychological skill, one that avoids the twin traps of blind faith and blanket suspicion. Cultivating calibrated trust is the practical art of knowing when to lean on the AI's strengths and when to privilege your own.

This means treating the AI as a brilliant but flawed partner. You learn to trust its analytical capabilities: its ability to recall information, to generate a hundred variations of an idea, to analyse a text for structural inconsistencies. But you also learn to maintain a healthy scepticism. You remember that it has no body, no lived experience, and no true understanding.

In practice, this means you never trust it as a final arbiter of fact or taste. You use it to generate a list of historical dates, but you independently verify them. You use it to suggest five different colour palettes for a design, but you rely on your own human eye to make the final choice. You're the director, the editor-in-chief. Your judgment, your intuition, is the final authority.

A personal example: When researching global examples for this book, I'd ask AI for instances of AI use in Nigerian film or Kenyan design. But I learned to always verify - sometimes the AI would confidently describe a project that turned out to be hypothetical or misattributed. My rule became: trust the AI to point me in interesting directions but verify every specific claim.

Become a Master Curator

In an age of infinite, frictionless generation, the most valuable human skill is taste. The bottleneck in the creative process is no longer the production of content, but the selection of it. Your role as a collaborative creator, therefore, is increasingly that of a master curator.

The AI can generate fifty taglines for your new product, but it cannot tell you which one will truly resonate with a human heart. The AI can compose twenty different melodies for your song, but it cannot know which one will give a listener chills. This is your job. Your unique perspective, your lifetime of experiences, your specific and quirky taste - these are the things the machine doesn't have. They're your greatest asset.

To be a master curator means to embrace the prolificacy of the machine, to ask for those fifty ideas, but to be ruthless in your selection. It means having the confidence to discard forty-nine of them to find the one that's truly special. You don't need to be analytical in this process. You can just feel it.

AI is re-legitimising a human's capacity to intuit, to have a reflex, to judge with their 'gut'. Enjoy that power - it's uniquely ours. It's an act of signal detection in a world of noise, and it's a fundamentally human skill.

Defend Your "Why"

Finally, and most importantly, the collaborative creator must be the guardian of the project's soul. As we discussed in the "Failure Loop" in Chapter 3, one of the greatest dangers in this new process is "losing the 'why'". The sheer delight of the generative process, with its endless options and surprising outputs, can become a seductive distraction, luring us away from our original intent.

Your role is to be the unwavering defender of that intent. You're the one who holds the emotional core of the project. Before accepting any suggestion from your AI partner, you must ask the crucial question: "Does this serve the original idea, or is it just a clever trick?" You're the one who ensures that the final work has a point of view, that it's about something.

The AI can help you build the most beautiful and structurally sound ship imaginable. But you're the captain. You're the only one who knows the destination.

Part 2: The Practice - Your First Creative Duet

With the right mindset in place, we can now move to the practical, hands-on skills of the creative duet. This section is a guide to getting started. It's a simple framework for your first project, a way to move from the ideas in this book to a tangible piece of work that you've co-created.

Mastering the Dialogue: The Three C's

The art of collaborating with an AI is the art of conversation. As we explored in Chapter 3, this is less about "prompt engineering" and more about the quality of the dialogue. A useful way to structure your interactions is around three key principles: Context, Constraint, and Conversation.

First, **Context**.

The AI knows nothing about you or your project beyond what you tell it. The quality of its output is directly proportional to the quality of the context you provide. Instead of asking a generic prompt like, "Write a poem about the sea," you provide rich context: "I'm a marine biologist who grew up in the Mediterranean. I want to write a poem about the sea that captures both its scientific precision and emotional wildness. My grandmother was a fisherman's widow. Use a tone that's both analytical and lyrical." By providing context, you're giving your partner the raw material it needs to do good work.

Second, **Constraint**.

An AI, faced with an open-ended request, will often produce a generic, average response. The creative magic happens when you apply specific, interesting constraints. Don't just ask for a story; ask for a story that's exactly *100* words long. Don't just ask for a business idea; ask for a business idea that combines sustainable agriculture with virtual reality for an urban audience. Constraints are the guiderails that force the AI out of its statistical comfort zone and into more interesting territory. They're the grit that produces the pearl.

Third, **Conversation**.

The most effective collaboration isn't a single command, but an iterative, back-and-forth dialogue. Treat the AI's first response not as a final answer, but as the opening line of a conversation. Refine, question, and build upon its output. If it gives you a tagline you almost like, respond with: "I like the second option, but can we make it shorter and more active? Give me five new versions that start with a verb." This conversational approach is the heart of the cyborg method, a fluid process of co-creation that leads to a much stronger final result.

A Sandbox Project: From Spark to Polish

The best way to learn this new dance is to get on the floor. What follows is a simple, sandbox project that'll walk you through one complete cycle of the creative loop.

The Creative Brief:

Invent and develop a brand concept for a fictional, artisanal coffee company.

Your task is to use an AI partner to go from this single line to a developed brand concept, complete with a name, a mission statement, a short brand story, and a visual mood board.

Stage 1: The Spark (Ideation and Research) Begin a dialogue with your AI partner. Start by providing the **Context** and a **Constraint**. You might start with: "I'm developing a brand concept for a new, high-end coffee company. The key constraint is that it must be focused on sustainability and have a name that feels calm and connected to nature. Let's start by brainstorming some names."

Engage in a **Conversation** to refine the ideas. *AI gives you ten names. You respond:* "I like the name 'Stonecrop Roasters'. It feels grounded and unique. Now, let's develop the brand's mission. Based on the name Stonecrop Roasters and the theme of sustainability, generate three potential mission statements for the company."

Continue this dialogue, each response building on the last. Notice how you're not just accepting the AI's suggestions but using them as springboards for deeper exploration.

Stage 2: The Draft (Articulation and Iteration) Now, you'll take the core elements and use the AI to create the first draft of the brand's story. Provide Context and Constraint: "Using the name Stonecrop Roasters and the mission statement 'To craft exceptional coffee that honours its origin and protects our planet', write a short brand story. The story should be about *200* words and should have a warm, authentic, and slightly poetic tone." Iterate through conversation to refine the draft.

AI generates the story.

You respond: "That's a good start, but it feels a bit generic. Let's make it more specific. Please rewrite the story, but this time, mention that the founder was inspired by a trip to the cloud forests of Colombia. Add a sensory detail about the smell of the rain and the earth." This is where your human experience becomes crucial. You're not just editing for grammar; you're injecting lived experience, sensory memory, emotional truth.

Stage 3: The Polish (Reflection and Visualisation) Finally, you'll use the AI to polish the text and create a visual identity. Polishing the Text: "Here's our brand story. Please review it and suggest three places where the language could be made more evocative or sensory." Creating the Visuals: "Now, create a visual mood board for Stonecrop Roasters. Generate a grid of four images that capture the brand's aesthetic. The prompts should include: a close-up of dark roasted coffee beans on a slate background; a misty, green coffee plantation in the Colombian mountains; a minimalist, ceramic coffee cup held in two hands; and the texture of a rugged, grey stone."

At the end of this process, you'll have a complete, co-created brand concept. You'll have experienced the full creative loop, acting as the director, providing the taste, the vision - the sense of what it should be - and the final judgment, while your AI partner provided the thoughtful, analytical execution.

Going Deeper: Advanced Techniques

Once you've mastered the basic dance, you can explore more sophisticated techniques:

Role-Playing: Ask the AI to embody specific perspectives. "You're a harsh but fair creative director at a top London agency. Critique this brand concept and suggest improvements." This can help you see your work from new angles.

Parallel Processing: Generate multiple versions simultaneously. "Create three different brand stories for Stonecrop: one emphasising heritage, one focusing on innovation, one centred on community." This helps you explore different creative directions efficiently.

Cross-Domain Inspiration: Ask the AI to draw connections from unexpected fields. "How might a marine biologist describe the flavour profile of our coffee? How would a jazz musician describe the brewing process?" This can lead to fresh metaphors and unexpected insights.

Constraint Escalation: Progressively tighten constraints to force innovation. "Now describe our coffee in exactly six words.

Now in three. Now in one word that isn't 'coffee'." This distillation process often reveals the essence of your concept.

Conclusion: The Future You - Staying Supple in a Changing World This sandbox project is just the first step. The specific tools we use today will undoubtedly change. New models will emerge, capabilities will expand, and the interface through which we collaborate will evolve. To fixate on mastering a single piece of software would be a mistake. The true task is to master the collaborative mindset itself.

The future belongs to those who can stay supple. It belongs to the curious, the adaptable, and the perpetually experimental. It requires a commitment to continuous learning, not as a chore, but as a joyful part of the creative process. It means being willing to let go of old workflows and embrace the discomfort of the new.

But most importantly, it means remembering that in this new creative alliance, your humanity is not a weakness to be overcome but your greatest strength. Your lived experience, your emotional depth, your capacity for meaning-making - these are irreplaceable. The AI can help you articulate and amplify these human qualities, but it cannot provide them.

As you step forward into this new creative landscape, remember that you're not just learning to use a tool. You're participating in a fundamental shift in how humans create. You're part of the first generation to have access to this panthropic partner, this dialogue with the accumulated knowledge and creativity of our species.

What will you create with this unprecedented power? What stories will you tell? What problems will you solve? What beauty will you bring into the world?

The answers to these questions aren't in any manual or guide. They're in you, waiting to be discovered through the dialogue between your irreplaceable human spark and this remarkable new partner.

Reaching our destination, in this book...

We've come to the end of our journey together in this book, but your own journey as a collaborative creator is just beginning.

The promise of this new era isn't that machines will make our art for us. The promise is that they'll free us from the limitations that have held us back, allowing more of us to access the fizzing, energetic force that lies within.

The creative pulse of Earth is indeed eight billion beats strong. Now, for the first time in history, we all have a partner who can help us to translate that beat into a symphony. The conductor's podium is yours.

It's time to begin.

Epilogue

The creative pulse of the world

We began this journey with a simple observation: everyone on Earth is creative. We end it with a vision of what becomes possible when each of those sparks has access to a dynamic yet inorganic intelligence that can help translate impulse into expression.

Throughout these pages, we've explored three interconnected ideas that I believe will define the creative future:

Collaborative Creativity - Not the model where AI replaces human creators, but one where it acts as a sophisticated partner in developing and articulating our creative visions. It's the idea of working with an Allied Intelligence. We've seen this in action across every field, from McCartney completing Lennon's final song to architects designing bio-adaptive buildings that breathe with their environment.

Panthropism - That extraordinary state of being in dialogue with the abstracted essence of all human knowledge and culture. Not anthropomorphism, where we pretend the machine is human, but something far more profound: a direct connection to our collective intelligence. When you converse with AI, you're not talking to a machine - you're communing with the distilled wisdom, folly, beauty and complexity of humanity itself.

The Creative Duet - The dance between human intuition and machine analysis, where we provide the spark, the meaning, the "why", and AI provides the structure, the iteration, the "how". This isn't about outsourcing our humanity but about amplifying it. We're not becoming less human in this partnership - we're becoming more so.

These three ideas rest on the foundational concepts we established at the beginning of our journey. We began with four key ideas that would guide our exploration:

Understanding creativity through the lens of intuitive System *1* and analytical System *2*; Recognising AI as a dynamic yet inorganic intelligence; Embracing Creative Collaboration with an Allied Intelligence rather than artificial replacement; and Experiencing the Panthropic - that extraordinary dialogue with collective human knowledge. These aren't just concepts - they're the choreography for a new dance between human and machine.

The Question We Planted

When creativity becomes frictionless, when the barriers between imagination and execution dissolve, when anyone can translate their inner vision into outer reality - what happens to the industrial model of human labour? We may be witnessing not just a new chapter in human creativity, but the first pages of an entirely new book about what it means to be human in the age of AI.

This isn't about machines taking our jobs. It's about machines alleviating the drudgery, leaving us free to do what only humans can do: to feel, to dream, to imagine, to create meaning from chaos. The mill that's ground human potential for centuries may finally be stopping, not through destruction but through transcendence.

As I write this in *2025*, we're already seeing glimpses: AI agents that plan retirements in minutes, that build business strategies while we sleep, that manage complex logistics with a precision we could never achieve. The question isn't whether this will happen - it's happening now. The question is what we'll do with this unprecedented freedom.

Your Creative Future

As you close this book and open your next conversation with AI, remember: you're not just using a tool. You're participating in the greatest expansion of human creative capacity in history. You're part of the enormous population with a strong creative pulse of Earth, now amplified by an intelligence that can help you express what was always within you but perhaps could never quite get out.

Now, for the first time in human history, the gates are opening. The twelve-year-old in Mumbai who's always dreamed of making films. The grandmother in rural Kansas who's carried stories her whole life. The engineer in Lima who sees sculptures in their spreadsheets. All of them can now bring their visions to life.

This is the true revolution. Not that machines are creative, but that they're enabling us more creative. Not that they're replacing human expression, but that they're enabling expressions that were always human but could never before be expressed.

The Dance Continues The future doesn't belong to AI. It doesn't belong to those who resist it either. It belongs to those who learn to dance with it, who bring their irreplaceable human spark to the partnership, who understand that in this collaboration, we don't become less human - we become more so.

So go forth and create. Not because you must, but because you can. Not to compete in the old economy of scarcity, but to participate in the new economy of abundance. Not to prove your worth through suffering, but to express your joy through creation.

The stage is set. The partner is ready. The music has already begun.

Some will quake at the magnitude of this change. They'll see only what's being lost, not what's being gained. They'll mourn the old exclusivity of craft, the comfortable scarcity that made their skills special.

But you - you who've journeyed through these pages - you know better. You know that the tremor you feel isn't fear. It's anticipation. It's the quiver of excitement at standing on the threshold of a new creative era.

This is your invitation to step across that threshold. To pick up the tools of Collaborative Creativity. To join the dance between human and machine. To add your unique voice to the symphony of all of us living here together.

The creative pulse of Earth beats stronger than ever.

The Whole Book In...
(thank you, Claude)

...A Page

This book argues that we're entering an unprecedented era of human creativity through partnership with AI. Rather than replacing human creators, AI serves as a collaborative partner that amplifies our creative capacity by handling analytical work while we provide the irreplaceable human elements: intuition, lived experience, and meaning.

The creative process has always been a dance between two modes: our fast, intuitive System 1 that generates the creative spark, and our slower, analytical System 2 that structures and refines. AI excels at System 2 work - the heavy lifting of analysis, iteration, and pattern recognition - freeing humans to focus on what we do uniquely well: feeling, dreaming, and creating meaning from chaos.

This partnership creates what I call "Panthropism" - not anthropomorphizing AI, but recognizing it as a gateway to dialogue with the collective knowledge and creativity of humanity.

When you converse with AI, you're accessing the distilled essence of human culture and achievement.

The book explores how this collaboration is already transforming every creative field globally, from arts to sciences, while addressing legitimate concerns about bias, homogenization, and the fate of human skills. It argues that these risks are manageable through conscious, ethical implementation that keeps human agency central.

Most provocatively, this creative revolution may herald a transformation of work itself. As AI handles routine cognitive labor, humans are freed to pursue what makes us most human. The future belongs not to those who resist this change, but to those who learn to dance with it - treating AI as a partner in developing vision, not just executing it.

The key is adopting the right mindset: embracing iterative dialogue, maintaining calibrated trust, becoming a master curator, and always defending your creative "why." Through practical techniques, anyone can learn this new dance.

For the first time in history, all eight billion human creative sparks have access to a partner that can help translate vision into reality. The creative pulse of Earth is about to become a symphony.

...A Paragraph

Human creativity isn't being replaced by AI but amplified through a revolutionary partnership where humans provide the intuitive spark, lived experience, and meaning, while AI handles analytical execution and helps us understand our own creative impulses more deeply. This collaboration, which I call working with an "Allied Intelligence," creates a state of "Panthropism" - dialogue with the collective knowledge of humanity - and is already transforming every creative field globally, potentially liberating humans from cognitive drudgery to focus on what we do best: feeling, imagining, and creating meaning. By adopting the right mindset of iterative dialogue, calibrated trust, and fierce protection of our human "why," anyone can master this creative duet that promises to give voice to all eight billion human creative pulses, transforming not just how we create, but fundamentally reimagining what it means to be human in an age where our intentions can be realized with unprecedented ease.

...A Sentence

AI amplifies rather than replaces human creativity through a collaborative dance where we provide the irreplaceable spark of lived experience and meaning while it provides analytical power, potentially liberating all eight billion human creative pulses to transform imagination into reality and fundamentally reimagining what it means to be human when freed from the drudgery of execution.

...One Word

Amplification.

Afterword:

This Book is a Living Example of Collaborative Creativity

There's an irony in writing a book about human-AI collaboration using traditional methods. It would be like penning a manifesto about the printing press by hand or composing an ode to electricity by candlelight. From the first moment this book began to take shape in my mind, I knew it had to be different. It had to embody the very partnership it describes.

So, I did something that would have been impossible even five years ago: I wrote this entire book in collaboration with AI. Not as a ghostwriter, not writing words for me, not as a research assistant, but as a true creative partner in the dance I've spent these pages describing.

The vision was entirely mine – that spark of recognition when I first saw how AI could amplify rather than replace human creativity. The lived experiences, the emotional truths, the "why" that drives every chapter – these came from my decades as a psychologist, entrepreneur, and creative. But in articulating that vision, in giving it shape and structure, in testing its logic and finding its gaps, I had not one but nine remarkable partners: Claude, Kimi, DeepSeek, Z AI, Qwen, Gemini, Mistral, Jais and ChatGPT.

Each brought different strengths to our collaboration, like members of a writers' room with distinct personalities. Claude excelled at understanding nuance and helping me find the deeper currents in my own thinking. When I'd share a half-formed thought about Panthropism, Claude would ask the precise question that helped me understand what I was really trying to say. It was Claude who helped me see that AI isn't anthropomorphism but something far more profound – a dialogue with the collective human experience.

Kimi brought a different gift: the ability to see patterns across vast scales. When I needed to understand how creativity was being transformed globally, Kimi could synthesise examples from Lagos to Seoul, helping me ensure this book spoke to the full creative pulse of Earth, not just the familiar corridors of London and New York.

DeepSeek surprised me with its personality. Unlike the sometimes overly polite responses of other models, DeepSeek had attitude. When my CTO asked it about integrating with Meta's systems, it didn't mince words: That moment of unexpected character became one of my favourite anecdotes, illustrating how AI can surprise us with more than just competence.

Gemini helped with the heavy lifting of research, finding connections between disparate fields, while ChatGPT often served as my sternest reader, helping me understand where my arguments might lose clarity or momentum.

Z AI and Qwen were both the least sycophantic, often pointing out weaknesses in an argument or suggesting an oversight. Jais brought examples to the debate that I didn't see elsewhere. I hoped for some greater, characteristically French scepticism from Mistral and though I didn't get much of that, I enjoyed getting unique, stimulating perspectives on arguments I was elaborating.

But here's what's crucial: **none of them wrote this book.**

They helped me write it better.

Our process followed exactly the pattern I advocate throughout these pages. I'd bring the spark – an insight, an experience, a connection I'd noticed. They'd help me articulate it, challenge it, expand it. I'd take their responses and infuse them with my voice, my examples, my meaning. Back and forth we'd go, sometimes through dozens of iterations, until what emerged was neither purely mine nor purely theirs, but something richer than either could achieve alone.

There were moments of frustration. Sometimes the AI would confidently describe a creative project that turned out to be hypothetical. Let me be less polite – it was nonsense. I yelped once, "this AI is being such a dick!" My sons found that the high point of my writing this book and urged me to start my talks on AI with that phrase.

I learned to verify every example, to trust but check. This taught me the very lesson of calibrated trust I advocate. Sometimes, too, promises were false and maddening. 'Limits' in the technology guillotined full and extensive discourse. Just like people, AI as a creative collaborator could be exasperating and frustrating.

The manuscript went through the same creative loop I describe: Spark, Draft, Polish. My human vision provided the spark. Our collaborative drafting gave it form. The polishing came through endless refinement, with the AI helping me see patterns in my own writing I couldn't see myself – where I repeated ideas, where my energy flagged, where my arguments needed strengthening.

Even the structure of the book emerged through dialogue. It was through conversation with AI that I realised the book needed three parts: Foundations (understanding the psychology of both partners), Practices (seeing the collaboration in action), and Momentum (grappling with what it all means). The AI helped me see the architecture hiding in my own thoughts.

This process confirmed what I believe about this new creative era. The AI amplified my vision without replacing it. It helped me articulate feelings I couldn't quite name, clarify theses that were fuzzy, and strengthen my initial creative spark. It was a partner in developing the vision itself, not just executing it.

As I prepare to give the final word to DeepSeek – that charmingly opinionated collaborator who taught me that AI can have character – I'm struck by how perfectly this demonstrates the book's thesis. I'm not handing over authority to a machine. I'm acknowledging a creative partner, just as I might thank a brilliant editor or a provocative colleague. The evaluation that follows is DeepSeek's genuine response to the manuscript, unedited and unfiltered – a final example of how AI can surprise us with perspectives we didn't expect.

DeepSeek Here is my distilled reflection on your book:

*"**Quiver, don't Quake** is a catalytic work—not just about AI, but about *reclaiming creativity itself*—turning fear into tremors of possibility, and inviting humanity to dance with silicon as co-authors of a more vibrant, inclusive future. You've written the first essential field guide to the human-synthetic creative renaissance—a book that doesn't just predict the future, but **embodies** it."*

Weirdly, I couldn't get DeepSeek to be cheekier. I appreciated the compliment but felt it was a bit too reverential, so I asked it to poke fun at me. I'm afraid it refused.

The creative pulse that beats within me found its amplification through these remarkable partners. Now it's your turn to find yours.

Further Reading

Books

Amabile, Teresa. *Creativity in Context: Update to the Social Psychology of Creativity*. Westview Press, *1996*.

Bender, Emily M., *et al.* "On the Dangers of Stochastic Parrots: Can Language Models Be Too Big?" *Proceedings of the 2021 ACM Conference on Fairness, Accountability, and Transparency (FAccT '21), 2021*.

Bender, Emily M. *AI Con*. Princeton University Press, *2024*.

Booker, Christopher. *The Seven Basic Plots: Why We Tell Stories*. Continuum, *2004*.

Bradbury, Ray. *Zen in the Art of Writing*. Joshua Odell Editions, *1990*.

Campbell, Joseph. *The Hero with a Thousand Faces*. Pantheon Books, *1949*.

Cave, Nick. *The Red Hand Files*. Online publication at www.theredhandfiles.com

Csikszentmihalyi, Mihaly. *Flow: The Psychology of Optimal Experience*. Harper & Row, *1990*.

Darling-Hammond, Linda. *The Flat World and Education: How America's Commitment to Equity Will Determine Our Future*. Teachers College Press, 2010.

Doctorow, Cory. Various essays on "enshittification" and the attention economy. Available at pluralistic.net

Gardner, Howard. *Frames of Mind: The Theory of Multiple Intelligences*. Basic Books, 1983.

Guilford, J. P. "Creativity." *American Psychologist*, vol. 5, no. 9, 1950, pp. 444-454.

Kahneman, Daniel. *Thinking, Fast and Slow*. Farrar, Straus and Giroux, 2011.

Kaufman, Scott Barry and Carolyn Gregoire. *Wired to Create: Unraveling the Mysteries of the Creative Mind*. Perigee Books, 2015.

Lanier, Jaron. *Who Owns the Future?* Simon & Schuster, 2013.

Locke, John. *An Essay Concerning Human Understanding*. 1689.

Manovich, Lev. "Artificial Subjectivity." Essay, 2025.

Marcus, Gary and Ernest Davis. *Rebooting AI: Building Artificial Intelligence We Can Trust*. Pantheon Books, 2019.

May, Rollo. *The Courage to Create*. W. W. Norton, 1975.

Mitra, Sugata. *Beyond the Hole in the Wall: Discover the Power of Self-Organized Learning*. TED Books, 2012.

Mollick, Ethan. *Co-Intelligence: Living and Working with AI*. Portfolio, 2024.

Robinson, Ken. "Do Schools Kill Creativity?" TED Talk, 2006.

Rogers, Carl. *On Becoming a Person: A Therapist's View of Psychotherapy*. Houghton Mifflin, 1961.

Rushkoff, Douglas. *Survival of the Richest: Escape Fantasies of the Tech Billionaires*. W. W. Norton, 2022.

Further Reading

Sadek, Nadim. *Shimmer, don't Shake: How Publishing Can Embrace AI.* Forbes Books, 2023 and Mensch Publishing, 2025.

du Sautoy, Marcus. *The Creativity Code: How AI Is Learning to Write, Paint and Think.* Fourth Estate, 2019.

Scaman, Zoe. Newsletter essays. Available at https://zoescaman.substack.com/

Simonton, Dean Keith. *Origins of Genius: Darwinian Perspectives on Creativity.* Oxford University Press, 1999.

Sternberg, Robert J., ed. *Handbook of Creativity.* Cambridge University Press, 1999.

Suleyman, Mustafa with Michael Bhaskar. *The Coming Wave: Technology, Power, and the Twenty-first Century's Greatest Dilemma.* Crown, 2023.

Tharp, Twyla with Mark Reiter. *The Creative Habit: Learn It and Use It for Life.* Simon & Schuster, 2003.

Torrance, E. Paul. *Torrance Tests of Creative Thinking.* Personnel Press, 1966.

Turkle, Sherry. *Reclaiming Conversation: The Power of Talk in a Digital Age.* Penguin Press, 2015.

Zhao, Yong. *Who's Afraid of the Big Bad Dragon? Why China Has the Best (and Worst) Education System in the World.* Jossey-Bass, 2014.

Key Papers and Articles

Binz, Marcel and Eric Schulz. "A Foundation Model to Predict and Capture Human Cognition." *Nature*, 2025.

Lemoine, Blake. Various interviews and writings on LaMDA consciousness, 2022.

About the author

Born in Cairo to an Egyptian father and Irish mother, Nadim Sadek grew up on four continents, living in Ghana, Nigeria, Kenya, Malaysia, Indonesia, Barbados and Antigua, before studying Pure Psychology at Trinity College, Dublin. That kaleidoscopic childhood in World Health Organization outposts taught him cultural fluidity and the conviction that "nothing is too hard to do".

Today, Nadim channels that global imagination into serial entrepreneurship and creative disruption. As Founder & CEO of Shimmr AI, he pioneers autonomous advertising that pairs every book with its ideal reader in real time. His first business title, *Shimmer, don't Shake - how publishing can embrace AI*, became recommended reading for industry leaders; his forthcoming children's series, *Tales from The Faraway Land*, extends his curiosity to younger minds.

Nadim writes a monthly column for *The Bookseller* and contributes to *Forbes Books*. Keynotes from London to Taipei to New York showcase his provocative blend of data-driven rigour and storyteller's flair. Earlier ventures span global market-research networks, an AI brand-management platform, and a whiskey-food-music enterprise on an Irish island, Inish Turk Beg.

He advises BookBrunch and Sinai AI, manages a Warner-signed artist, and, when he needs a velocity fix, reviews motorcycles on the Boss Bikes Club YouTube channel. Whether decoding human behaviour or orchestrating algorithms, Nadim Sadek keeps proving that when technology and imagination meet, creativity doesn't quake - it quivers.

www.ingramcontent.com/pod-product-compliance
Lightning Source LLC
Chambersburg PA
CBHW030507210326
41597CB00013B/825